Communications
in Computer and Information Science **706**

Commenced Publication in 2007
Founding and Former Series Editors:
Alfredo Cuzzocrea, Dominik Ślęzak, and Xiaokang Yang

More information about this series at http://www.springer.com/series/7899

Leonid Kalinichenko · Sergei O. Kuznetsov
Yannis Manolopoulos (Eds.)

Data Analytics and Management in Data Intensive Domains

XVIII International Conference, DAMDID/RCDL 2016
Ershovo, Moscow, Russia, October 11–14, 2016
Revised Selected Papers

 Springer

Editors
Leonid Kalinichenko
Federal Research Center "Computer Science
  and Control" of RAS
Moscow
Russia

Sergei O. Kuznetsov
National Research University Higher School
  of Economics
Moscow
Russia

Yannis Manolopoulos
Aristotle University of Thessaloniki
Thessaloniki
Greece

ISSN 1865-0929   ISSN 1865-0937   (electronic)
Communications in Computer and Information Science
ISBN 978-3-319-57134-8   ISBN 978-3-319-57135-5   (eBook)
DOI 10.1007/978-3-319-57135-5

Library of Congress Control Number: 2017938325

This Springer imprint is published by Springer Nature
The registered company is Springer International Publishing AG
The registered company address is: Gewerbestrasse 11, 6330 Cham, Switzerland

Preface

In 2016 the XVIII International Conference on Data Analytics and Management in Data-Intensive Domains (DAMDID/RCDL 2016) was held during October 11–14 in the Holiday Center, Ershovo (Moscow region).

By tradition, the Data Analytics and Management in Data-Intensive Domains Conference (DAMDID) is planned as a multidisciplinary forum of researchers and practitioners from various domains of science and research, promoting cooperation and exchange of ideas in the area of data analysis and management in domains driven by data-intensive research. Approaches to data analysis and management being developed in specific data-intensive domains (DID) of X-informatics (such as X = astro, bio, chemo, geo, medicine, neuro, physics, etc.), social sciences, as well as in various branches of informatics, industry, new technologies, finance, and business contribute to the conference content.

The program of DAMDID/RCDL 2016 alongside with the traditional data management topics reflected a rapid move into the direction of data science and data-intensive analytics. Three invited plenary talks of the conference emphasized the key problems of methods and facilities of data analytics and management in DID. In the keynote of Andrey Rzhetsky (Professor of Genetic Medicine of Chicago University) that opened the conference, the approaches for studying the mechanisms of cancer were presented, which are being developed by the University of Chicago consortium in which Andrey Rzhetsky acts as the PI according to the DARPA Big Mechanism program (http://www.darpa.mil/program/big-mechanism). These approaches are centered on the development of contemporary methods of data analysis in DID including cognitive methods based on the understanding of the natural language resembling those of IBM Watson, but focused on the hypothesis formation based on the causal relationships detected in the texts, cancer mechanism modeling to automatically predict therapeutic clues, and application of robot scientist methods for the organization of experiments. The invited talk of Sophia Ananiadou, Director of the National Centre for Text Mining (NaCTeM) at Manchester University, opened the second day of the conference. NaCTeM acts as part of the University of Chicago consortium supported by DARPA. In this talk the approaches for automatic reconstruction of the pathway models are considered. These approaches are based on the discovery of the various kinds of relationships between the concepts of arbitrary nature in the texts. Finally, the program on the third day was opened with the keynote of Dimitrios Tzovaras (Director at the Information Technologies Institute of the Centre for Research and Technology Hellas in Thessaloniki) in which the analytical survey of the European strategy in the area of research infrastructures was presented.

The conference Program Committee reviewed 57 submissions and accepted 27 of them as full papers, 16 as short papers, three as posters, two as demos, while nine submissions were rejected. According to the conference program, these 43 oral presentations (of the full and short papers) were structured into 13 sessions including

Semantic Modeling in DID, Data Analysis Methods, Knowledge Management, Learning Management, Semantic Search and Navigation, Pattern Analysis in Recommender Systems, Research Data Infrastructures (in Astronomy, Astrophysics, Material Sciences, Earth Monitoring), Data Extraction from Texts, Data Integration and Sharing, and Text Analysis Systems. Most of the presentations showcased the results of research conducted in the organizations located in the territory of the Russian Federation including Dubna, Ekaterinburg, Irkutsk, Kazan, Moscow, Novosibirsk, Obninsk, Puschino, Tomsk, Tula, Saint Petersburg, Yaroslavl, and Zvenigorod.

All accepted papers are published in the local proceedings; all full papers are published in the CEUR Workshop volume 1752 (http://ceur-ws.org/Vol-1752). This CCIS volume is a post-conference proceedings containing 19 selected, revised, and extended papers. The process of preparing nine of them for this volume included translation from the Russian language into English. It is worth mentioning that Sophia Ananiadou kindly agreed to transform the extended abstract of her invited talk into a full paper.

The contents of the volume are structured into the following sections reflecting the wide spectrum of topics related to the DAMDID conferences: Semantic Modeling in Data-Intensive Domains, Knowledge and Learning Management, Text Mining, Data Infrastructures in Astrophysics, Data Analysis, and Research Infrastructures. The position paper by Yannis Manolopoulos is included separately, providing an introduction into bibliometrics (considering such related fields as scientometrics, citation analysis, journal impact factor) with a warning that the widely used metrics and rankings should be applied with great skepticism.

DAMDID/RCDL 2016 would not have been possible without the support of the Russian Foundation for Basic Research, the Federal Agency of Scientific Organizations of the Russian Federation, the National Research Nuclear University MEPhI, and the Federal Research Center Computer Science and Control of the Russian Academy of Sciences.

Finally, we thank Springer for publishing this proceedings volume, containing the invited and selected research papers, in their CCIS series, The Program Committee of the conference appreciates the possibility to use the Conference Management Toolkit (CMT) sponsored by Microsoft Research, which provided great support during various phases of the paper submission and reviewing process.

February 2017   Leonid A. Kalinichenko
                                                      Sergei O. Kuznetsov
                                                      Yannis Manolopoulos

Organization

General Chair

Igor Sokolov   Federal Research Center Computer Science and Control of RAS, Russia

Program Committee Co-chairs

Leonid Kalinichenko   Federal Research Center Computer Science and Control of RAS, Russia

Sergei O. Kuznetsov   National Research University Higher School of Economics, Moscow, Russia

Yannis Manolopoulos   Aristotle University of Thessaloniki, Greece

PhD Workshop Co-chairs

Ladjel Bellatreche   LIAS/ISAE-ENSMA, Poitier, France

Ilia Sochenkov   Federal Research Center Computer Science and Control of RAS, Russia

Organizing Committee Co-chairs

Victor Zakharov   Federal Research Center Computer Science and Control of RAS, Russia

Alexander Nevzorov   NRNU MEPhI, Russia

Boris Pozin   MIEM NRU HSE, Russia

Organizing Committee

Dmitry Briukhov   Federal Research Center Computer Science and Control of RAS, Russia

Nikolay Skvortsov   Federal Research Center Computer Science and Control of RAS, Russia

Dmitry Kovalev   Federal Research Center Computer Science and Control of RAS, Russia

Anna Nadezhdina   Federal Research Center Computer Science and Control of RAS, Russia

Evgenia Dudareva   Federal Research Center Computer Science and Control of RAS, Russia

Yulia Trusova   Federal Research Center Computer Science and Control of RAS, Russia

Tatjana Dudareva   Protey Travel, Russia
Sergey Vereshchagin   INASAN, Russia

Supporters

Russian Foundation for Basic Research
Federal Agency of Scientific Organizations of the Russian Federation
National Research Nuclear University MEPhI
National Research University Higher School of Economics (NRU HSE)
Federal Research Center Computer Science and Control of the Russian Academy of
Sciences (FRC CSC RAS)
Moscow ACM SIGMOD Chapter

Coordinating Committee

Igor Sokolov   Federal Research Center Computer Science and Control of
                              RAS, Russia (Co-chair)
Nikolay Kolchanov   Institute of Cytology and Genetics, SB RAS, Novosibirsk,
                              Russia (Co-chair)
Leonid Kalinichenko   Federal Research Center Computer Science and Control of
                              RAS, Russia (Deputy Chair)
Arkady Avramenko   Pushchino Radio Astronomy Observatory, RAS, Russia
Pavel Braslavsky   Ural Federal University, SKB Kontur, Russia
Vasily Bunakov   Science and Technology Facilities Council, Harwell,
                              Oxfordshire, UK
Alexander Elizarov   Kazan (Volga Region) Federal University, Russia
Alexander Fazliev   Institute of Atmospheric Optics, RAS, Siberian Branch,
                              Russia
Alexei Klimentov   Brookhaven National Laboratory, USA
Mikhail Kogalovsky   Market Economy Institute, RAS, Russia
Vladimir Korenkov   JINR, Dubna, Russia
Mikhail Kuzminski   Institute of Organic Chemistry, RAS, Russia
Sergey Kuznetsov   Institute for System Programming, RAS, Russia
Vladimir Litvine   Evogh Inc., California, USA
Archil Maysuradze   Moscow State University, Russia
Oleg Malkov   Institute of Astronomy, RAS, Russia
Alexander Marchuk   Institute of Informatics Systems, RAS, Siberian Branch,
                              Russia
Igor Nekrestjanov   Verizon Corporation, USA
Boris Novikov   St. Petersburg State University, Russia
Nikolay Podkolodny   ICaG, SB RAS, Novosibirsk, Russia
Aleksey Pozanenko   Space Research Institute, RAS, Russia
Vladimir Serebryakov   Computing Center of RAS, Russia
Yury Smetanin   Russian Foundation for Basic Research, Moscow
Vladimir Smirnov   Yaroslavl State University, Russia

Sergey Stupnikov Federal Research Center Computer Science and Control of RAS, Russia
Konstantin Vorontsov Moscow State University, Russia
Viacheslav Wolfengagen National Research Nuclear University MEPhI, Russia

Victor Zakharov Federal Research Center Computer Science and Control of RAS, Russia

Program Committee

Karl Aberer EPFL, Lausanne, Switzerland
Plamen Angelov Lancaster University, UK
Arkady Avramenko Pushchino Observatory, Russia
Ladjel Bellatreche LIAS/ISAE-ENSMA, Poitiers, France
Pavel Braslavski Ural Federal University, Yekaterinburg, Russia
Vasily Bunakov Science and Technology Facilities Council, Harwell, UK
Panos Chrysanthis University of Pittsburgh, USA
Paul Cohen University of Arizona, USA
Ernesto Damiani Khalifa University, Abu Dhabi, United Arab Emirates
Yuri Demchenko University of Amsterdam, The Netherlands
Boris Dobrov Research Computing Center of MSU, Russia
Shlomi Dolev DCS, Ben-Gurion University of the Negev, Beer-Sheva, Israel
Alexander Elizarov Kazan Federal University, Russia
Alexander Fazliev Institute of Atmospheric Optics, SB RAS, Russia
Olga Gorchinskaya FORS, Moscow, Russia
Evgeny Gordov Institute of Monitoring of Climatic and Ecological Systems SB RAS, Russia
Maxim Gubin Google Inc., USA
Ralf Hofestadt University of Bielefeld, Germany
Leonid Kalinichenko FRC CSC RAS, Moscow, Russia
George Karypis University of Minnesota, Minneapolis, USA
Nadezhda Kiselyova IMET RAS, Russia
Alexei Klimentov Brookhaven National Laboratory, USA
Mikhail Kogalovsky Market Economy Institute, RAS, Russia
Vladimir Korenkov Joint Institute for Nuclear Research, Dubna, Russia
Efim Kudashev Space Research Institute, RAS, Russia
Dmitry Lande Institute for Information Recording, NASU, Russia
Vladimir Litvine Evogh Inc., California, USA
Giuseppe Longo University of Naples Federico II, Italy
Natalia Loukachevitch Moscow State University, Russia
Ivan Lukovic University of Novi Sad, Serbia
Oleg Malkov Institute of Astronomy, RAS, Russia
Yannis Manolopoulos School of Informatics of the Aristotle University of Thessaloniki, Greece

Alexander Marchuk	A.P. Ershov Institute of Informatics Systems SB RAS, Russia
Archil Maysuradze	Moscow State University, Russia
Piotr Mirowski	Google Deep Mind, London, UK
Igor Nekrestyanov	Verizon Corporation, USA
Boris Novikov	Saint Petersburg State University, Russia
Gennady Ososkov	Joint Institute for Nuclear Research, Russia
Dmitry Paley	Yaroslav State University, Russia
Stelios Paparizos	Google, San Francisco, USA
Milan Petkovic	Technical University of Eindhoven, The Netherlands
Nikolay Podkolodny	Institute of Cytology and Genetics SB RAS, Russia
Jaroslav Pokorny	Karlov University in Prague, Czech Republic
Natalia Ponomareva	Scientific Center of Neurology of RAMS, Russia
Alexey Pozanenko	Space Research Institute, RAS, Russia
Boris Pozin	EC-Leasing, Russia
Andreas Rauber	Vienna TU, Austria
Alexander Rogov	Petrozavodsk State University, Russia
Timos Sellis	RMIT, Australia
Vladimir Serebryakov	Computing Centre of RAS, Russia
Nikolay Skvortsov	FRC CSC RAS, Russia
Vladimir Smirnov	Yaroslavl State University, Russia
Leonid Sokolinskiy	South Ural State University, Russia
Valery Sokolov	Yaroslavl State University, Russia
Sergey Stupnikov	FRC CSC RAS, Russia
Alexander Sychev	Voronezh State University, Russia
Bernhard Thalheim	University of Kiel, Germany
Dmitry Tsarkov	Manchester University, UK
Alexey Ushakov	University of California, Santa Barbara, USA
Natalia Vassilieva	Hewlett-Packard, Russia
Konstantin Vorontsov	MSU, Russia
Alexey Vovchenko	FRC CSC RAS, Moscow, Russia
Domagoj Vrgoc	Center for Semantic Web Research, Chile
Peter Wittenburg	MPI for Psycholinguistics, Germany
Viacheslav Wolfengagen	MEPhI, Russia
Yury Zagorulko	Institute of Informatics Systems, SB RAS, Russia
Victor Zakharov	FRC CSC RAS, Russia
Oleg Zhizhimov	Institute of Computing Technologies of SB RAS, Russia
Sergey Znamensky	Institute of Program Systems, RAS, Russia

Contents

Data Analysis

Research Infrastructures

Position Paper

Semantic Modeling in Data Intensive Domains

Conceptualization of Methods and Experiments in Data Intensive Research Domains

Nikolay A. Skvortsov$^{(\boxtimes)}$, Leonid A. Kalinichenko,
and Dmitry Yu Kovalev

Federal Research Center "Computer Science
and Control" of Russian Academy of Sciences, Moscow, Russia
nskv@mail.ru, leonidandk@gmail.com,
dm.kovalev@gmail.com

Abstract. Nowadays research of various scopes especially in natural sciences requires manipulation of big volumes of data generated by observation, experiments and modeling. Organization of data-intensive research assumes definition of domain specifications including concepts (specified by ontologies) and formal representation of data describing domain objects and their behavior (using conceptual schemes), shared and maintained by communities working in the respective domains. Research infrastructures are based on domain specifications and provide methods applied to such specifications, collected and developed by research communities. Tools for organizing experiments in research infrastructures are also supported by conceptual specifications of measuring and investigating object properties, applying the research methods, describing and testing the hypotheses. Astronomy as a sample data intensive domain is chosen to demonstrate building of conceptual specifications and usage of them for data analysis.

Keywords: Conceptual modeling · Ontology · Conceptual schema · FAIR data principles · Research infrastructure · Research experiment

1 Introduction

Research process is critically dependent on growing and complementing massive data collections obtained from observations, experiments and modeling. At the same time the quality and hence depth of required data analysis are growing. An approach to research which supposes selection of data sources and formulation of problems in their terms becomes difficult to apply in case of a multitude of heterogeneous data sources and ways to analyze data from them. Programs for data analysis and problem solving dependent on specific data sources do not provide scalability for the heterogeneous and massive data sources, interoperability and reuse of data analysis methods developed in plenty of different researches [13].

A focus is shifted from searching and linking of data sources for research towards analysis of massive data collections for discovering new knowledge in the research subject domains [10]. Scientific methods are developed for generalization and

© Springer International Publishing AG 2017
L. Kalinichenko et al. (Eds.): DAMDID/RCDL 2016, CCIS 706, pp. 3–17, 2017.
DOI: 10.1007/978-3-319-57135-5_1

classification, discovering and analysis of entities and phenomena of interest, evaluation of physical parameters of objects, generation and testing of scientific hypotheses, specialized procedures used in certain research domains. These methods can be automatically applied over data of massive collections, and access to them is provided for communities working in the research infrastructures.

For comprehensive investigation of the specific real-world entities, it is important to share data, tools, results, methods and specifications defining the semantics of entities and phenomena in the domain as well as the semantics of methods applied to them.

In recent years, requirements to curation of research data, metadata, and other published research information resources are actively discussed. The following principles are agreed for the stored and published data: findability, accessibility, interoperability and reusability (FAIR). Adherence to those principles is intended to make data reusable for both human and machine. Communities of researchers and vendors of analytical tools, research instruments and data owners are interested in the shared access to heterogeneous data collections.

The principle of findability is implemented with using global identification, describing data with metadata and indexing for search. The principle of availability provides guaranteed extraction data by identifiers using common protocols and formats with authorization and authentication if required. The interoperability of data resources implies usage of presentations of data and metadata accepted by communities and taking into account knowledge of the domain and providing semantic interpretation by machine. The reusability requires compliance with the requirements and standards of communities, clear licensing of access to data, change tracking and versioning, technical environment for reproducing results [32].

There are special requirements for the citations of data, including long-term storage after data life, citing of researcher names, information about cited pieces, versions, time [27].

Informality of FAIR principles rises a number of different interpretations, including application of Linked Data principles to provide FAIR ones [33], or lists of more detailed informal requirements based on FAIR [2], or just simplified numerical rating of conformity with FAIR principles [14].

Analysis of projects cited as samples of FAIR principles implementation shows that most of them do not consider anything more than data and its metadata ignoring human and machine understandability, curation of methods, services, workflow with their metadata. Some of the projects consider identification and metadata related to whole collections in a data bank, and do not provide specifications and direct access to entities or records in collections, do not provide provenance of data. Some of them do not provide for the integration of resources in homogeneous manner to access them.

Due to the above, one of the objectives of the research communities is conceptualization of a research subject domain in order to develop the specifications properly defining the shared data, methods and processes. To provide FAIR principles and to support science with scientific methods and tools applicable to domain objects, we propose an adequate approach to conceptualization of subject domains. It is based on an explicit description of the semantics of entities, methods and processes during the statement of scientific problems and development of algorithms to solve them, providing compliance of problem descriptions with the semantic specifications of the domain. On the other hand, various data sources are semantically mapped to conceptual specifications of the domain. Analytical tasks are formulated in terms of conceptual specifications of the domain and solved using the relevant data and methods.

In this paper we show how the conceptual specifications of domains supported by scientific communities can be used to organize research in the FAIR way. Conceptual specifications in astronomy are used to demonstrate the approach. The next section is devoted to principles of definition of subject domain specifications for research. Section 3 describes approaches to collecting scientific methods and prospects for research infrastructures based on scientific method collections. Section 4 is devoted to usage of subject domain specifications for organization of experiments, description and testing scientific hypotheses, development of workflows in the research infrastructures by their communities to manipulate data and methods in experiments.

The paper advances the investigation of [25] with the analysis of requirements to principles of interoperability and reusability of all kinds of the research information objects including data sources, knowledge of a subject domain, information structures and methods related to domain objects, programs, research instruments and workflows.

2 Specification of Subject Domains

General concern of the FAIR principles are problems of reusability of data, research results, developed methods and algorithms used in the research processes to make data and research instruments accessible, sharable and reusable for researchers.

Collaborative usage of resources in research requires global identification and description of information objects for matching of objects regardless of their sources. There are many standard systems of global identification, in particular, DOI, URI, and PURL. Using one of them with any data object or other used in research process is enough to identify them uniquely and to organize access by identifiers.

Metadata specifications are supporting information accompanying resources and describing their semantics or own properties. Possible metadata models are usually based on text descriptions, standardized sets of attributes, lists of keywords, or thesauri and ontologies. On the other hand, in accordance with the FAIR principles metadata should be understandable both for human and machine to select relevant resources and automatically solve problems.

To be understandable by machine a system should be formal. It is not enough just to search similar objects and propose them to users. Automation implies taking decisions by machines (however human should have this possibility too). Thus, full formal specifications of the used resources and evidence-based inference over metadata are the best way to provide machine-understandable data.

Specificity of different domains of research should be reflected in the vocabularies developed for the metadata of objects. Formal ontologies provide well-defined vocabularies including concepts and relations of domains and bind them to the theories for restriction of their interpretations by machines. So the conceptualization of domains applying formal ontologies is assumed to be an appropriate approach to specify research resources with metadata.

Organization of metadata registries makes resources findable. Usually registries contain the metadata related to data collections as a whole or the metadata describing structures of data collections. However not only data should be findable in the research infrastructures, but every kind of the information objects that can be addressed during research in the subject domains. Those are data, metadata, publications, implementations of research methods, workflows describing the research processes. Thus, no matter which kind of object is used for research, it should be supplied with metadata in terms of ontologies. Inference over ontologies lets to select them from collections and access by selected identifiers.

The process of domain conceptualization primarily implies development of ontologies by research communities for formalization and systematization of the knowledge on general entities and phenomena of the subject domains. Community members operate within the ontological commitment defined by shared ontologies, i.e., use of the concepts of the subject domain in a consistent way with respect to the theories specified by the ontologies. Ontologies are important for the automation of consistency control on any manipulations with the domain concepts. An interaction of communities in solving interdisciplinary problems requires simultaneous querying using different domain vocabularies. In that case, the researchers should commit to the specifications of several domains.

The knowledge related to any kind of objects appearing in the definitions of a subject domain includes the following:

- concepts of entities appearing in the domain, investigated ones or related to them;
- concepts defining characteristics and behavior of domain objects;
- concepts describing scientific methods and correlations existing in the domain;
- concepts defining approaches to observation of objects, modeling of domain entities, scientific experiments.

Languages for formal ontology representation operate on concepts, relations and constraints usually expressed in a subset of predicate logic, a descriptive logic or another formal models. Ontologies preferably do not use scalar data types since on the ontological level any relation would better be described explicitly to provide an ability of unambiguous interpretation of concepts. Also ontologies traditionally do not use specifications of the methods. However, the definitions related to an entity behavior should be specified in ontologies too. For this purpose, the concepts related to correlations of entity characteristics and processes are defined.

Besides definition of ontologies, the conceptualization process also includes development of conceptual schemes of the subject domains. The difference of a conceptual scheme from an ontology is the foremost one in its purpose and application [19]. The conceptual scheme defines not just concepts, but a presentation of information on the domain objects and behavior specifications to specify their state changing and manipulating. However, if ontologies are defined, then the conceptual schemes should be developed in accordance with the knowledge defined in these ontologies. Principles of development of the conceptual schemes of the subject domains using ontological knowledge are described in [30].

Languages for a conceptual scheme specification include definition of the abstract data types that represent states of objects characterized by a set of attributes values of which should conform to the defined data types from simple scalar to the object and association ones. Constraints define consistency of the object states. Characteristics of types and attributes can be defined by metadata. Classes are defined as the sets of objects of the same type. Behavior of the domain objects is expressed by the type methods.

It is advisable to accompany elements of conceptual schema specifications and instance objects with metadata annotating them by related ontological concepts in order to define their semantics. If the conceptual schemes involve ontological knowledge, then information on elements of the schemes, which were formed in accordance with the concepts of ontologies, is stored as metadata. Thus, semantics of scheme elements are explicitly specified in terms of the domain concepts.

Resources containing data and services for a research community of a certain domain, in their turn, are organized and systematized in accordance with the onto-logical concepts too. Formalization of specifications of ontologies and conceptual schemes is critically important to provide for the semantic integration of information resources and reproducibility of programs over the domain specifications. Proof-based approach is intended to distinguish a possibility of automatic reasoning applying the formal domain specifications from non-formal conceptualization making possible only manual provisions for decisions required in such cases as, e.g., integration (or inter-operability) of heterogeneous resources. Formal reasoning in the conceptual approach is required in many cases, including such as:

- verification of consistency of ontologies and conceptual schemes of the subject domain;
- control of interoperability of combined or substituted specifications;
- verification of the conceptual scheme specifications (being under development) for compliance with the knowledge reflected in ontologies;
- searching for information resource specifications semantically relevant to domain specifications;
- verification of the used information resources for compliance with the domain specifications.

Ability to apply automatic reasoning depends on the language of specification. The problems mentioned above are solvable for specifications reduced to descriptive logics. In particular, the descriptive logics are used for the OWL ontology representation language dialects.

For ability to use inference not only on the ontological level but over data too, it is preferable to specify conceptual schemes using unified logical models, e.g., in par-ticular dialects of RIF language [7]. RIF + OWL allow expressing rules over specifi-cations in OWL and defining formal specifications of behavior of domain objects and algorithms for solving problems directly over a scheme defining structures and con-straints in OWL. In multi-dialect architecture, reasoning over specifications is per-formed using different provers depending on the chosen RIF dialects [18].

Both OWL and RIF languages support identification of the elements of specifi-cations with the URI global identifiers, and annotations to define binding of elements with metadata in terms of domain ontologies in OWL [26, 29].

For languages reduced to the first-order logic, the same problems can be solved interactively (e.g., using inference of program specification refinement [11]). A refining specification can substitute the refined one. In particular, during the development of conceptual schemes based on ontologies it is required that the conceptual scheme specification refines the ontological ones. To prove this, specifications of ontologies and conceptual schemes should be mapped to abstract machines notation (AMN) of B system, that provides a proof of refinement [23, 24, 28]. Concepts defining correlations and processes are mapped to operations, which should be refined by operations of the conceptual scheme types. Preconditions and post-conditions of operations should refine constraints of those concepts.

Ontologies and conceptual schemes of the subject domains are developed and maintained by communities working in those domains. So the specifications become important for the needs of research groups of communities. E.g., development of conceptual specifications in communities might be stimulated by a necessity to reach a semantic interoperability (or integration) of interacting components (or data collections), reuse of data resources and program reproducibility due to binding to semantics of the subject domains. Therefore, both the ontological level and the level of conceptual schemes should meet high requirements for completeness and formality of the specifications.

Existing domain specifications and standards accepted by wide communities of researches should be taken into account too. For example, in astronomy, in frame of the International Virtual Observatory Alliance (IVOA) the appropriate standards have been developed. Known ontologies [3, 5] are based on thesauri, do not contain essential concepts and relationships required by the researchers in their work, do not include various constraints of the domain or descriptions of behavior of objects, phenomena and observation processes. No formal ontologies intended for automatic reasoning are applied.

Another kinds of conceptual specifications developed in astronomy are the standards of conceptual schemes of most general domains referred to in almost every astronomical project. Such standards are:

- Space-Time Coordinate Metadata [9] is a scheme with parameters of various coordinate systems;
- Photometry Data Model [4] is a standard describing photometric systems, magnitudes, calibration functions and conversion between different systems;
- VOEvent [8] is a scheme describing observation of astrophysical objects and phenomena, including identification of objects and observers, coordinates, epochs and instruments of observation.

Schemes as well as ontologies in IVOA mostly do not include consistency constraints and behavior specifications, and do not define objects and methods of more specific domains. Conceptualization is necessary not only in such general domains within astronomy as astrometry, photometry and spectroscopy, but in the domains of interest of narrower and more specialized communities, as well as in overlapping domains, in which a cooperation of research teams and reuse of specifications often occurs. The specific research objects, methods and experiments should be defined.

Specifications below are fragments of the prototype of ontology in selected areas of astronomy developed by the authors of this paper. The modular structure of the ontology is required for the description of the specific domains within astronomy. It includes domains defined by specific research objects, observation and modeling methods, as well as approaches to research process.

The ontological specifications of scientific experiments have been developed. They include basic interdisciplinary concepts for research:

- an ontology of characteristics of measurements related to research objects, defining concepts for units of measurement, errors, distribution laws and others;
- an ontology of measurement correlations as a conceptual framework for ontological specifications of object behavior, including the concept of correlation and such subconcepts as function, method, hypothesis, law and others.

Definitions of several modules of conceptual specifications in the OWL language [6] intended for the scientific experiment description framework are presented. The module of concepts related to the *ofMeasurement* parameters of objects includes the *Measurement* concept with the relation *isMeasurementOf* binding it to an astronomical object, concepts of parameter values, units of measurement, measurement accuracy, characterized by statistical and systematic errors.

```
Class (Measurement
  restriction (hasValue
    allValuesFrom (Value))
  restriction (hasUnit
    allValuesFrom (MeasurementUnit))
  restriction (hasError
    allValuesFrom (MeasurementError) maxCardinality (1))
  restriction (isMeasurementOf
    allValuesFrom (AstrObject) maxCardinality (1)))
Class (MeasurementUnit
  restriction (hasScaleFactor
    allValuesFrom (ScaleFactor) maxCardinality (1))
  restriction (hasProjection
    allValuesFrom (ScaleProjection) maxCardinality (1)))
Class (MeasurementError
  restriction (isErrorOf
    allValuesFrom (Measurement) maxCardinality (1)))
Class (StatisticalError partial MeasurementError)
Class (SystematicError partial MeasurementError)
Class (Value
  restriction (isValueOf
    allValuesFrom (Measurement) maxCardinality (1)))
ObjectProperty (isValueOf
  domain (Value) range (Measurement)
  inverseOf (hasValue))
ObjectProperty (isMeasurementOf
  domain (Measurement) range (AstrObject)
  inverseOf (hasMeasurement))
ObjectProperty (isErrorOf
  domain (MeasurementError) range (Measurement)
  inverseOf (hasError))
```

Another module defines concepts related to common characteristics of all astronomical objects. Specification of the astronomical object concept (AstrObject) includes relations to other concepts of ontology: coordinates, measurements of various kinds of observed and astrophysical parameters, belonging to compound objects, and others:

```
Class (AstrObject
  restriction (hasIdentifier
    allValuesFrom (Identifier))
  restriction (hasCoordinate
    allValuesFrom (Coordinate))
  restriction (inEpoch
    allValuesFrom (Epoch) maxCardinality (1))
  restriction (hasMeasurement
    allValuesFrom (Measurement))
  restriction (hasMorphology
    allValuesFrom (Morphology) maxCardinality (1))
  restriction (hasProcess
    allValuesFrom (Process))
  restriction (isComponentOf
    allValuesFrom (CompoundObject)))
```

Specific concepts of measurements are defined as subconcepts of *Measurement* concepts in specialized modules ontology in which they are used. The module of the astronomical objects contains the general astrophysical parameters of objects, such as temperature, mass, diameter, luminosity. In particular, mass is a common feature of all astronomical objects:

```
Class (Mass partial Measurement)
```

Ontological module describing the domain of stars includes the concept of stellar object (*StellarObject*) as an entity in the Galaxy, single one or a component of a compound, *Star* concept as a single stellar object, multiple star concept (*MultipleStar*) as a stellar object consisting of components, and a number of specific characteristics of stars:

```
Class (StellarObject partial AstrObject)
Class (Star partial StellarObject)
```

In this module, we can define the stellar mass concept (*StellarMass*), which is a subconcept of *Mass*. It uses concepts of other modules limiting the type of entities described as the stars and determining as a unit mass of the Sun.

```
Class (StellarMass partial Mass
  restriction (isParameterOf
    allValuesFrom (StellarObject))
  restriction (hasUnit
    hasValue (SunMass)))
```

Developing of a conceptual scheme using ontological knowledge it is necessary to create structures for the representation of information (metadata) related to the domain objects. In the multidialect architecture [18] the different specification languages can be used for that. We use specifications in the SYNTHESIS language [17] for implementation of conceptual specifications in subject mediators [15]. Specifications in the OWL language and the RIF BLD rule language dialect has been mapped to the SYNTHESIS language [16]. Specifications in SYNTHESIS also has been mapped to the AMN notation for proving of the specification refinement [28].

An example of a conceptual schema specification in SYNTHESIS built in accordance with the ontological specifications defines a structure for the star objects with the attributes representing coordinates, stellar mass and metadata that defines the unit of measurement in masses of the Sun:

```
{ Star;
  in: type;
  crd: Coordinate;
  mass: Float;
  metaslot
    in: measurement;
    hasUnit: SunMass;
  end
}
```

Specifications of conceptual scheme also require the definitions of methods and functions. These specifications are generated on the base of ontological concepts that describe the correlations of object characteristics and processes. These issues are discussed in the next section.

3   Organization of Collections of Scientific Data and Methods

In the specialized domains of research defined for the research infrastructures the libraries of methods are actively developed today as well as the data collections for the research communities. They provide various kinds of services, instruments descriptions and facilities for the search of the data resources. A brief look at the examples of several infrastructures applied in astronomy for different purposes is provided below.

One of the infrastructures provided instruments for researcher collaboration in astronomy was the AstroGrid [1]. This is a grid infrastructure designed as the virtual observatory that consists of a set of nodes to support services, data or computing resources. AstroGrid had a registry as a collection of metadata describing registered resources with a limited set of attributes. It allowed searching for data sources and services supported on the available nodes.

WF4Ever project [12] aims at preserving workflows and results of data intensive researches for their reuse. As a part of this project, service libraries [31] are developed in various domains. They provide access to the existing web services and data sources, transformations of different data representation, support of various data manipulation. These services can be used as the elements of workflows. For astronomers these libraries were recognized by the scientific community primarily due to the simplicity of the workflow construction.

The pan-European Data Infrastructure (EUDAT) [22] is a network of collaborating, cooperating centers, combining the richness of numerous community specific data repositories with the permanence and persistence of some of Europe's largest scientific data centers. Service providers, both generic and thematic, and research communities have joined forces as part of a common framework for developing an interoperable layer of common data services. Known as the EUDAT Collaborative Data Infrastructure (CDI), this is essentially a European e-infrastructure of integrated data services and resources to support research. Hierarchies of terms referring to research disciplines, research methods and objects are developed by research communities. IVOA is represented in the infrastructure as one of research domains.

In the near future astronomical data volumes will be enriched with new sources. Projects such as wide-angle telescope LSST and the space observatory Gaia will generate streaming observation. To process such data the channels are prepared for their localization, access issues for various research institutions are discussed, and instruments of processing are developed [20].

For effective interaction within the community and for avoiding of independent overlapping works, collaboration of researchers should be based on the shared development and use of another kind of information, such as, e.g., the research plans, open collections of methods, algorithms, analysis tools, as well as the collaboration infrastructures ready to be used over largescale data sets. Conceptualization of the domain within the community and semantic approaches allow to systematize these necessary components of research infrastructures. Domain methods as well as the research data are bound with the conceptual specifications of the domain.

In every area of knowledge, domain-specific patterns and methods are collected for analysis of the respective types of entities. As well, a wide range of analytical methods and general tools used in any domain should be accessible to researchers.

In the ontological models that traditionally do not include elements for method specification, a conceptualization of entity behavior can be defined by concepts reflecting correlating characteristics and processes. E.g., one of the modules of the ontology might be a module defining interdependence of measurements. The *Correlation* concept implies the mutual dependency of two or more measured parameters of the domain entities:

```
Class (Correlation
  restriction (isCorrelationOf
    allValuesFrom (Measurement))
  restriction (hasRegression
    allValuesFrom (RegressionFunction))
  restriction (hasRMSDeviation
    allValuesFrom (RMSDeviation))
  restriction (isCausal
    allValuesFrom (TruthValue)))
Class (Hypothesis partial Correlation
  restriction (explains
    allValuesFrom (Phenomenon))
  restriction (derivedFrom
    allValuesFrom (Hypothesis))
  restriction (competesWith
    allValuesFrom (Hypothesis))
  restriction (hasProbability
    allValuesFrom (Probability))
  restriction (hasPValue
    allValuesFrom (Probability))
  restriction (hasQuality
    allValuesFrom (TruthValue)))
Class (Law partial Hypothesis
  restriction (hasQuality hasValue (True)))
```

At the ontological level concepts declaratively describe the research methods, hypotheses, laws, models, processes, experiments related to characteristics of the domain objects. For example, the hypothesis concept can be defined as a kind of statistically verifiable correlation. For statistical testing of the hypotheses, modeling can be used to provide their mathematical description. On the other hand, if possible, the experiments can be used to compare the results of modeling with a data of observations of real-world entities. The law concept can be defined as a hypothesis qualified to true under all possible tests.

Concepts of methods, hypotheses, laws are defined as subconcepts of the measurement correlations with the respective constraints of the dependent values. Their ontological descriptions are independent of specific implementations and representations of correlation such as tables with coefficients, mathematical formulas, distribution functions, programs, and other possible ways to describe them.

Let us consider a conceptualization of the initial mass function (IMF) of hypothesis defined within the Besancon Galaxy model [21] in terms of the domain ontologies described above. This hypothesis is based on the assumption of a quite constant distribution of stars with the different masses in a limited space of the Galaxy. In other words, the hypothesis implies a dependency of a number of stars on their masses in a fixed volume of space:

```
Class (InitialMassFunction partial Hypothesis
   restriction (isCorrelationOf
     ObjectSomeValuesFrom (StarMass))
   restriction (isCorrelationOf
     ObjectSomeValuesFrom (
       intersectionOf (
         Quantity
         restriction (hasElement
         allValuesFrom (Star))))))
```

Using the ontology, collections of different implementations of methods, hypotheses or laws can be systematized by various criteria related to the ontological concepts: by research objects, by characteristics of entities, by known methods and hypotheses, and others. Any such criterion can be used by researchers to search for the implementation of the existing methods for reuse.

Existing implementations of the general-purpose methods such as machine learning or numerical analytics can be used to design the experiments. However, it is preferable to wrap their usage in implementations the conceptual schema methods to define semantics of the action.

4   Specification of Research Experiments

Experiments over data in the research infrastructure are constructed using shared and interoperable data, services and workflows. Research experiments often include data analysis, construction of research hypotheses, modeling in accordance with hypotheses and testing models by observational data.

Besides providing access to data and method implementation collections, research infrastructures should include instruments for experiment supporting, in particular, formulation and testing of hypotheses. Conceptual specifications give advantages as for it.

Conceptual schemes should include methods and rules built in accordance with the concepts of methods, hypotheses, and laws of the subject domain. E.g., using ontological knowledge on the hypothesis of the initial mass function, specifications for modeling and testing this hypothesis should be created in the conceptual scheme. The type IMF in the SYNTHESIS language includes definition of the function signatures with the parameters corresponding to the relations in the *InitialMassFunction* correlation concept.

```
{ IMF;
  in: type;
  supertype: Hypothesis;
  draw_mass: {in: function;
  params: {+mass/Real, -quantity/Real}};
}
```

The function defines correlation of masses and quantities of stars. The abstract data type *IMF* presented above can be implemented in different ways to organize testing the initial mass function hypothesis. Existing implementations of models for the hypothesis can be found in method collections by ontological metadata. Subtypes of the *IMS* type are defined for specific implementations of models. New models can be implemented too.

On the other hand, the same specification of the *IMF* type is used to test hypotheses with observational data. Relevant data can be found and obtained from multiple catalogs integrated into the conceptual scheme of the domain. A subtype of the *IMF* type is constructed for this purpose. The hypothesis testing is performed by comparing the results of modeling to the observational data.

For modeling and testing hypotheses the workflows are developed over conceptual schemes of the subject domains. Workflows should implement experiment process including modeling, accessing to observational data, testing the model by comparison to observed data, and adjustment of the hypothesis parameters to generate and new models.

5 Conclusion

Efficiency of research conducted by the domain communities depends on availability of observational data, implementation of methods and models, and experiment planning which takes into account ontological knowledge related to entities, processes and dependencies of measurements.

In the paper, the issues of domain conceptualization for organizing the data intensive research are considered. Development of the research infrastructures based on formal conceptual specifications maintained by the domain research communities has some advantages. They provide for organizing collections of data, methods, and programs and search for them in terms of the domain ontologies. In particular, they provide for resolving the issues of interoperability (integration) of data collections and method implementations, formulation of research problems in terms of conceptual schemes of domains, as well as for the representation of results in structures known to the whole communities for simplification of reuse. They provide also for the development of specific instruments for the research experiments support, in particular, for testing hypotheses, for improving the reliability of the research results due to the semantic approaches and formal consistency verification.

The intention of this paper is to warn that the genuine support of the FAIR principles should provide facilities for various kinds of formal reasoning, some examples of which were presented in this paper. The first implementations of the FAIR data environment (e.g., [33]) do not exhibit any intention to reach the required completeness. Serious further researches are required, including systematic clarification and formalization of the FAIR data principles.

Acknowledgements. The work was partly supported by the Russian Foundation for Basic Research grants 15-29-06045, 16-07-01028, and 16-07-01162.

References

1. AstroGrid. http://www.astrogrid.org/
2. Guidelines on FAIR Data Management in Horizon 2020: Directorate-General for Research and Innovation European Commission (2016). http://ec.europa.eu/research/participants/data/ref/h2020/grants_manual/hi/oa_pilot/h2020-hi-oa-data-mgt_en.pdf
3. IVOAO Ontology: University of Maryland (2010). http://www.astro.umd.edu/~eshaya/astroonto/
4. IVOA Photometry Data Model: Version 1.0. IVOA (2013). http://www.ivoa.net/documents/PHOTDM/
5. Ontology of Astronomical Object Types: Version 1.20. IVOA (2009). http://www.ivoa.net/documents/latest/AstrObjectOntology.html
6. OWL 2 Web Ontology Language: Document Overview, 2nd edn. W3C (2012). http://www.w3.org/TR/owl-overview/
7. Kifer, M.: Rule interchange format: the framework. In: Calvanese, D., Lausen, G. (eds.) RR 2008. LNCS, vol. 5341, pp. 1–11. Springer, Heidelberg (2008). doi:10.1007/978-3-540-88737-9_1
8. Sky Event Reporting Metadata (VOEvent): Version 2.0. IVOA (2011). http://www.ivoa.net/Documents/VOEvent/
9. Space-Time Coordinate Metadata for the Virtual Observatory Version 1.33: IVOA (2011). http://www.ivoa.net/documents/latest/STC.html
10. Hey, T., et al. (eds.): The Fourth Paradigm: Data-Intensive Scientific Discovery. Microsoft Research, Redmond (2009)
11. Abrial, J.R.: The B Book. Assigning Programs to Meanings. Cambridge University Press, Cambridge (1996)
12. Belhajjame, K., et al.: Workflow-centric research objects: a first class citizen in the scholarly discourse. In: ESWC2012 Workshop on the Future of Scholarly Communication in the Semantic Web (SePublica2012), pp. 1–12, Heraklion (2012)
13. Briukhov, D.O., Vovchenko, A.E., Kalinichenko L.A.: Support of the workflow specifications reuse by ensuring its independence of the specific data collections and services. In: Proceedings of the 15th Russian Conference on Digital Libraries RCDL 2013, vol. 1108, pp. 61–69. CEUR Workshop Proceedings (2013). (In Russian)
14. Doorn, P., Dillo, I.: FAIR data in trustworthy data repositories. DANS/EUDAT/OpenAIRE Webinar (2016). https://eudat.eu/events/webinar/fair-data-in-trustworthy-data-repositories-webinar
15. Kalinichenko, L.A., Briukhov, D.O., Martynov, D.O., Skvortsov, N.A., Stupnikov, S.A.: Mediation framework for enterprise information system infrastructures. In: Proceedings of the 9th International Conference on Enterprise Information Systems, ICEIS 2007. Volume Databases and Information Systems Integration, Funchal, pp. 246–251(2007)
16. Kalinichenko, L.A., Stupnikov, S.A.: OWL as yet another data model to be integrated. In: Proceedings of the 15th East-European Conference Advances in Databases and Information Systems, vol. 2, pp. 178–189. Austrian Computer Society, Vienna (2011)
17. Kalinichenko, L.A., Stupnikov, S.A., Martynov, D.O.: SYNTHESIS: a language for canonical information modeling and mediator definition for problem solving in heterogeneous information resource environments, 171 p. IPI RAS, Moscow (2007)
18. Kalinichenko, L., Stupnikov, S., Vovchenko, A., Kovalev, D.: Rule-based multidialect infrastructure for conceptual problem solving over heterogeneous distributed information resources. In: Catania, B., et al. (eds.) New Trends in Databases and Information Systems. AISC, vol. 241, pp. 61–68. Springer, Cham (2013)

19. Kogalovskiy, M.R., Kalinichenko, L.A.: Conceptual modeling in database technologies and ontological models. In: Kalinichenko, L.A. (ed.) Proceedings of the Workshop on Ontological Modeling, pp. 114–148. IPI RAS, Moscow (2008). (In Russian)
20. Luric, M., Tysoc, T.: LSST data management: entering the era of petascale optical astronomy. Proc. Int. Astron. Union **10**(H16), 675–676 (2012)
21. Robin, A.C., Reylé, C., Derrière, S., Picaud, S.: A synthetic view on structure and evolution of the Milky Way. Astron. Astrophys. **409**, 523–540 (2003). doi:10.1051/0004-6361: 20031117. Astrophysics Data System
22. Schentz, H., le Franc, Y.: Building a semantic repository using B2SHARE. In: EUDAT 3rd Conference (2014)
23. Skvortsov, N.A.: Application of concept refinement in salvation of ontology manipulation tasks. In: Proceedings of the Ninth Russian Conference on Digital Libraries RCDL 2007, pp. 225–229. Pereslavl University, Pereslavl-Zalesskij (2007). (In Russian)
24. Skvortsov, N.A.: Using of an interactive proving system for ontology mapping. In: Proceedings of the Eighth Russian Conference on Digital Libraries, RCDL 2006, Suzdal, pp. 65–69. P.G. Demidov Yaroslavl State University, Yaroslavl (2006). (In Russian)
25. Skvortsov, N.A., Kalinichenko, L.A., Kovalev, D.Y.: Conceptual modeling of subject domains in data intensive research. In: Selected Papers of the XVIII International Conference on Data Analytics and Management in Data Intensive Domains (DAMDID/RCDL 2016), vol. 1752, pp. 7–15. CEUR Workshop Proceedings (2016). (In Russian)
26. Skvortsov, N.A., et al.: Metadata model for semantic search for rule-based workflow implementations. Syst. Means. Inform. **24**(4), 4–28 (2014). IPI RAS, Moscow (In Russian)
27. Starr, J., et al.: Achieving human and machine accessibility of cited data in scholarly publications. PeerJ Comput. Sci. **1**, e1 (2015)
28. Stupnikov, S.A.: Mapping of canonical model core specifications in abstract machine Notation. In: Formal Methods and Models for Compositional Infrastructures of Distributed Information Systems: The Systems and Means of Informatics, pp. 69–95. IPI RAS, Moscow (2005). Special Issue (In Russian)
29. Tejo-Alonso, C., Berrueta, D., Polo, L., Fernández, S.: Metadata for web ontologies and rules: current practices and perspectives. In: García-Barriocanal, E., Cebeci, Z., Okur, M.C., Öztürk, A. (eds.) MTSR 2011. CCIS, vol. 240, pp. 56–67. Springer, Heidelberg (2011). doi:10.1007/978-3-642-24731-6_6
30. Vovchenko, A.E., et al.: From specifications of requirements to conceptual schema. In: Proceedings of the 12th Russian Conference on Digital Libraries RCDL 2010, pp. 375–381. Kazan Federal University, Kazan (2010). (In Russian)
31. Walton, N.A., et al.: Taverna and workflows in the virtual observatory. In: Astronomical Data Analysis Software and Systems ASP Conference Series, vol. 394, p. 309 (2007)
32. Wilkinson M., et al: The FAIR guiding principles for scientific data management and stewardship. Sci. data **3** (2016)
33. Wilkinson, M.D., et al.: Interoperability and FAIRness through a novel combination of web technologies. PeerJ Prepr. **5**, e2522v2 (2017). https://doi.org/10.7287/peerj.preprints.2522v2

Semantic Search in a Personal Digital Library

Dmitriy Malakhov[1(✉)], Yuri Sidorenko[1], Olga Ataeva[2],
and Vladimir Serebryakov[2]

[1] Lomonosov Moscow State University, Moscow, Russia
mda.develop@gmail.com
[2] Dorodnicyn Computing Centre of RAS, Moscow, Russia

Abstract. The article offers a solution to the problem of semantic search in a personal digital LibMeta library. It also describes the L-tag-based semantic search model. The article provides an algorithm to build up a keywords and clusters hierarchy by the means of iterative clustering and keywords extraction. This hierarchy is used to generate abstracts and extract L-tags from texts.

Keywords: Semantic search · Information search · Digital library · LibMeta · Clustering · Keywords extraction · L-tag

1 Introduction

It is traditionally believed that digital library resources consist of bibliographic records of traditional libraries and digital copies of the documents described by these records. Still technological development redefines the concept of both the libraries and their resources (that go beyond bibliographical records and their digital representation) and pushes the semantics of these resources to the forefront, which may call for various library resources classifications to be implemented. There have been developed different application field indexes that allow us to more specifically define resource topics. As a rule these means are either not enough to describe semantics, or are not sufficient to the new rules of library resources description, which leads to both more complex descriptions and more efforts taken to introduce new description means that correspond to the current demands.

With the new facilities provided by developing technologies a library user can make better use of all the digital library resources means. He or she has the ability to describe their field of interest in the standard terms of the subject field with dictionary thesauruses and ontologies at hand. This allows a user to organize and describe both his or her own collections and resources, as well as to make more detailed resource and field descriptions thus defining their terms more specifically.

The LibMeta [1] personal open semantic digital library is characterized by a flexible store of metadata for its own resources and types of described information resources. This way to describe the resources of the library makes the descriptions of its resources and objects universal and unattached to any subject field or users' field of interest. The description structure ensures maintained connections between different types of the recourses.

L. Kalinichenko et al. (Eds.): DAMDID/RCDL 2016, CCIS 706, pp. 18–30, 2017.
DOI: 10.1007/978-3-319-57135-5_2

The flexibility of the resources description is insured by the use of OWL ontologies for metadata storage. This format provides a number of benefits, such as:

- the ability to perform SPARQL queries;
- getting additional knowledge with logical inference;
- easier integration with other libraries;
- the ability to change schemes to match new requirements.

Semantic search is the search of documents by their contents. The LibMeta library allows to conduct semantic search on metadata with the use of SPARQL queries, still it doesn't allow to conduct semantic search on book texts. The ultimate goal of this work is to boost the quality of LibMeta library cervices with the help of the ability to conduct semantic search on the library book texts.

Thus it is necessary to realize the semantic search system for the LibMeta library book texts. This system should use a search query in a natural language, taking semantics into account, to look for the relevant book texts. The semantics is supposed to be supported by synonym and hyponym dictionaries.

This article develops the research published in CEUR [2]. The third part dedicated to the semantic search model has been replaced: instead of the S-tag model description this article introduces the L-tag semantic search model, which is based on new principals. Part 8 that deals with the semantic search system architecture has been reworked as well.

2   The Semantic Search Organization

There are different ways to organize semantic search on texts. The latest years have seen semantic texts annotation gaining more popularity. There exist a variety of ways to solve the issue of semantic annotation. Each of them focuses on attaching a set of semantically close tags to a document or a part of a document. These tags help you to find these documents later on. Besides you can look for documents using the regular full-text search and then take these tags into account while working with the document, thus obtaining more information [3]. It is common to use names, places, organizations names or other subjects as the tags [4].

RDF stores that contain a set of concepts and connections between them are often used to describe the tags. Some methods use information from Wikipedia as from a large source of knowledge as well [5]. The semantic annotation methods has lately been more and more addressing the use of the massive interconnected Linked Open Data cloud [4, 6]. For example, The National Archives of Great Britain (42 TB) has been annotated with the use of GATE annotation tool [7].

Semantic annotation is not the only way to organized search. There exist findings based on classic full-text search, which broaden it, adding the ability to use synonyms within the query. Thus there had been made up an ontology based on terms from various articles with the help of UDC [8], which was then used to broaden a user's query. The approach that deals with the information on syntax, morphology and punctuation appears to be of special interest as well [9]. Unfortunately, the approaches described above haven't been implemented and are not commonly used.

Let us then focus on an alternative way to organize semantic search with the use of synonym and hyponym dictionaries.

There have been conducted a lot of experiments on using synonym and hyponym dictionaries to improve the full-text search quality. It is widely known that the usage of synonyms and hyponyms makes searching results excessive and less accurate [10].

What makes this approach special is that only the meaningful parts of a text are indexed. Depending on the objective, these parts can feature paragraphs, sentences, word-combinations or a man-made tags (for example: a hashtag). Changing the size of the meaningful part lets us control the accuracy and fullness. For example, when the fullness is small and we see sentences being indexed, we can try to index combinations of sentences.

3   The L-tag-based Semantic Search

3.1   The L-tag

Suppose we are given the following elements:

- A set of documents D, where a document stands for whatever exists.
- An alphabet A, where an alphabet stands for any finite nonempty set.
- A set of terms T, where a term stands for an ordered multiset of elements from A.
- A set of L-tags LT, where an L-tag stands for an ordered multiset of elements from T.

For example: an alphabet can stand for any alphabet of a natural language. Words and word-combinations of the chosen natural language alphabet are the terms of this alphabet. A search query composed of a sequence of words and word-combinations, made of the natural language alphabet A is an L-tag on a T-set, where the T-set is a multiset of words and word-combinations in the natural language of the alphabet A.

Let us consider that each L-tag represents informational requirements and can be used by the user to represent this requirements. Let us also suggest that the users of the system will not represent different informational requirements using the same L-tag.

Informational requirements stand for a weighty set of documents which is necessary for the user to solve a problem, where the weight of a document is defined by its relevance for the problem-solving process. The ultimate goal of the information search is the assessment of the document weight within the informational requirements represented by the query. The definition suggests that the informational requirements can cross, be embedded or coincide. This feature will be used further to define the similarity function.

We are usually certain to know which documents are featured in the informational requirements described by an L-tag. Further in the text there will be introduced the semantic function and the similarity function, which are meant to value the informational requirements described by different L-tags.

3.2   The Similarity Function

We can define the similarity function as the following:

$$sim\!:\!\left(lt_1, lt_2\right) \rightarrow R \tag{1}$$

where lt_1, lt_2 are L-tags and R is a set of real numbers.

The similarity function (1) should satisfy the following rules:

1. The similarity function (1) is always greater than or equal to zero. Besides the function is not limited.
2. If the informational requirements of the L-tag lt_1 does not fully include the informational requirements of the L-tag lt_2 than the similarity function equals zero.
3. The more the informational requirements of the L-tag lt_1 include the informational requirements of the L-tag lt_2, the bigger is the similarity function (1).

The ultimate goal of the similarity function (1) is to determine how much the informational requirements of one L-tag include the informational requirements of the other L-tag. Based on the similarity function (1) estimation we can make conclusions about the interrelationships between the informational requirements described by two L-tags.

It is obvious that the similarity function is not generally symmetric, as it does not imply that when the informational requirements of one L-tag include the informational requirements of the other L-tag it stays all the same the other way round.

Let us note that the similarity function (1) sets the L-tags hierarchy, based on the informational requirements hierarchy. The similarity function can be differently estimated depending on the understanding of the way the informational requirements are represented by an L-tag. It is important that the similarity function (1) estimations are consistent with the rules stated above.

3.3 The Similarity Function Estimation Example

Let us consider the following L-tags:

- l_1: 'A paper'.
- l_2: 'A scientific paper'.
- l_3: 'A scientific paper on computer science'.
- l_4: 'A scientific paper on physics'.

We can say that the informational requirements of the L-tag l_1 include the informational requirements of the L-tag l_2, which include the informational requirements of both the L-tag l_3 and the L-tag l_4.

According to the similarity function (1) rules, it should possess the following properties:

$$sim\!\left(l_1, l_1\right) \geq sim\!\left(l_1, l_2\right) \tag{2}$$

$$sim\!\left(l_1, l_2\right) \geq sim\!\left(l_1, l_3\right) \tag{3}$$

$$sim\!\left(l_1, l_2\right) \geq sim\!\left(l_1, l_4\right) \tag{4}$$

$$sim(l_2, l_1) = 0 \tag{5}$$

Property (2)–(4) follow the second similarity function (1) rule, when property (5) follows the first rule.

Let us note that the similarity function (1) rules provide no regulations for the value of the function for the (l_2, l_3) and (l_2, l_4) pairs:

$$sim(l_2, l_3) \ ? \ sim(l_2, l_4) \tag{6}$$

For example, we can estimate the similarity function (1) the following way:

$$sim(l_1, l_1) = sim(l_2, l_2) = sim(l_3, l_3) = sim(l_4, l_4) = 1 \tag{7}$$

$$sim(l_1, l_2) = 0.6 \tag{8}$$

$$sim(l_2, l_3) = 0.7 \tag{9}$$

$$sim(l_2, l_4) = 0.6 \tag{10}$$

$$sim(l_1, l_3) = sim(l_1, l_2) * sim(l_2, l_3) = 0.42 \tag{11}$$

$$sim(l_1, l_4) = sim(l_1, l_2) * sim(l_2, l_4) = 0.36 \tag{12}$$

In any other cases the similarity function (1) estimation equals zero.

3.4 The Semantic Function

The semantic function can be defined as follows:

$$sem:(d, lt) \rightarrow R \tag{13}$$

where d is a document, lt is an L-tag and R is a set of real numbers.

The semantics function (13) should satisfy the following rules:

1. The semantic function (13) is always greater than or equal to zero. Besides the function is not limited.
2. The semantic function (13) is equal to zero if the informational requirements of the L-tag lt are not satisfied with the document d.
3. The more is the weight of the document d in the informational requirements of the L-tag lt, the bigger is the semantic function (13).

The ultimate goal of the semantic function (13) is to define to determine how much the informational requirements of an L-tag correspond to the information in the document. Based on the semantic function (13) estimation we can make conclusions about the informational requirements described by an L-tag.

The semantic function (13) can be differently estimated depending on the understanding of the way the informational requirements are represented by an L-tag and of the way this or that information from the document can satisfy this or that informational requirement. The semantic function (13) estimation should be consistent with the rules stated above.

3.5   The Semantic Function Estimation Example

Let us consider the following L-tags:

- l_1: 'A paper'.
- l_2: 'A scientific paper'.
- l_3: 'A scientific paper on computer science'.
- l_4: 'A scientific paper on physics'.

Let us consider the following documents:

- d_1: carries the information of the document being a paper.
- d_2: carries the information of the document being a scientific paper.
- d_3: carries the information of the document being a scientific paper on computer science.
- d_4: carries the information of the document being a scientific paper on physics.

For example, we can estimate the semantic function (13) the following way:

$$sem(d_1, l_1) = sem(d_2, l_2) = sem(d_3, l_3) = sem(d_4, l_4) = 1 \tag{14}$$

$$sem(d_2, l_1) = 0.6 \tag{15}$$

$$sem(d_4, l_1) = 0.36 \tag{16}$$

$$sem(d_3, l_1) = 0.42 \tag{17}$$

$$sem(d_3, l_2) = 0.7 \tag{18}$$

$$sem(d_4, l_2) = 0.6 \tag{19}$$

In any other cases the semantic function (13) estimation equals zero.

3.6   The Semantic Function Interpolation

As seen from the previous examples of the similarity function (1) estimation and the semantic function (13) estimation, the latter can be excessive (15–19).

Suppose we are given a set of points $(d, lt) \in P$, where d is a document, lt is an L-tag. Suppose the semantic function (13) is known in the points of P and is unknown in the other points, when the similarity function (1) is known everywhere. Then we can interpolate the semantic function (13):

$$
\begin{cases}
sem(d, lt), (d, lt) \in P \\
\max_{(d, lt_1) \in P} \left(sem\left(d, lt_1\right) * sim\left(lt_1, lt\right)\right), (d, lt) \notin P
\end{cases}
\tag{20}
$$

Interpolation usually means calculating a value of a function in a point based on the value of the function in the vicinity of this point. In our case to estimate the semantic function (13) we use the values of the semantic function (13) in the vicinity of the point. This point should on the one hand be sufficiently close to that point, and on the other hand have a sufficiently big value of the semantic function (13). Interpolation can as well be conducted differently in case it satisfies the semantic function (13) rules.

The semantic function (13) interpolation helps to make its estimation process easier. Instead of estimating the semantics function (13) in all points (14–19) it is enough to estimate it in the points (14). For the cases (15–19) the estimation can be obtained through interpolation (20).

Thus, with the similarity function (1) estimation for all the L-tags and with the semantics function (13) estimation for P, we can estimate the weight of a document in the informational requirements of any L-tag.

It is crucial to understand that the more points belong to the set P, the more precise the semantics function (13) estimation is. This is why the size of the set P is a compromise between the difficulty and the quality of the semantics function (13) estimation.

3.7   The Semantic Search Model

Let us consider that an L-tag lt describes a document d, if:

$$
sem(d, lt) > 0
\tag{21}
$$

Suppose we are given:

- A set of L-tags LT.
- A set of documents D, described by the L-tags from LT.
- The similarity function (1) on the data set LT.
- A set of points $(d, lt) \in P$, where $d \in D, lt \in LT$.
- The semantic function (13) on the data set P.
- The semantic function (13) interpolation on the data set D and LT.

The semantic search model can be defined as implementation of the information search model:

- Each document should be described by a set of L-tags. Suppose an L-tag lt describes a set of document d, if the value of the semantic function (13) for the pair (d, lt) equals zero.
- A query can feature a set of L-tags LT that expresses the informational requirements of a user.

- The relevance function fixed on a data set of documents D and a data set of queries LT can be estimated by the semantic function interpolation (20).

Thus, the suggested model epitomizes the information search concept. What makes this model specific is that it implies singling out the main and most meaningful parts of a document, expressed by L-tags, and estimating the value of these L-tags for the document with the help of semantic function interpolation (20). Besides, the similarity function (1) estimation can be done with the use of synonyms from a thesaurus, which enables us to compare L-tags both syntactically and semantically. All this lets us claim this model to be the semantic search model.

Using the model enables us to reduce the semantic search objective to the objective of defining similarity function (1) on the data set LT and to the objective of defining the semantic function (13) on the data set P. This lets us single out or assign the meaning-reflecting L-tags beforehand by calculating the semantic function (13) and then use these L-tags when searching.

Obtaining the relevant documents data set $\{d\} \subseteq D$ for a certain query $q \in LT$ can feature three stages:

4. Obtaining a data set of L-tags $\{lt\} \subseteq LT$:

$$sim(q, lt) > 0 \tag{22}$$

5. Obtaining a set of documents $\{d\} \subseteq D$:

$$sem(d, lt) > 0, lt \in \{lt\} \tag{23}$$

6. Sorting the set of documents $\{d\}$ by their meanings:

$$sem(d, q), d \in \{d\} \tag{24}$$

4   Extracting L-tags from Texts

As shown above, we can carry out semantic search using L-tags if they are extracted from book texts and had their semantic function (13) estimated. Let us now consider the way to automatically extract L-tags from texts.

Text keywords stand for words and word-combinations that convey the main idea of the text and differ it from the other texts of the collection.

To match the content of a text, an L-tag should include keywords of this text. An L-tag can be a keyword, a sentence or a paragraph that includes a keyword. In this case the value of the semantic function (13) should depend on the L-tag keywords significance. Thus the objective to extract L-tags can be reduced to keyword text scanning.

As a keyword of a text should separate this text from the other texts, keywords depend on both the text they belong to and the whole collection of texts which opposes that text. If we divide a set of texts into semantically close text groups, we can see a keyword as a marking indicator that characterizes a text and the corresponding text group.

From another point of view, if we use keywords as indicators to divide texts, we can boost the speed and quality of such division by cutting indicators space and filtering noise.

Thus, clustering as the process of dividing texts into groups can use marked keywords, when the keywords marking algorithm can use the results of clustering. With clustering it is possible to divide a great multitude of texts into groups (clusters) and then look for keywords for the included texts according to those groups. This process can be repeated several times, alternating clustering with keywords extracting.

As a result we get a hierarchical structure of the documents in the collection and the corresponding keywords hierarchy.

5   Extracting Keywords

Suppose we are given a data set of texts D on the data set of terms T. A term stands for a word or a word-combination. The data set D is divided into a number of clusters C. The objective is to scan the data set D and then extract the keywords that characterize the cluster of this text.

Keyword marking traditionally involves two stages. The first stage features marking keywords candidates. At this stage stop words are deleted, parts of speech or the candidates that are not featured in Wikipedia headlines are filtered. The second stage is meant to check how close the candidates are to the text semantic-wise. This check is conducted with the help of supervised and unsupervised machine learning algorithms [11].

What makes our objective differ from the standard keywords extraction objective is the fact that the keywords should depend on both the document under scanning and some other documents that belong to the same field. The proposed approach can be upgraded with the help of the standard problem-solving algorithms [11].

Let us consider a simple way to solve the problem. Suppose we have a random text $d \in D$. For every cluster $c \in C$ and term $t \in T$. Let us estimate the probability of $d \in c$ when $t \in d$:

$$P(d \in c | t \in d) = \frac{|(t \in d \wedge d \in c)|}{|(t \in d)|} \tag{25}$$

where

- $|(t \in d \wedge d \in c)|$ is the number of documents from the cluster that feature the term t.
- $|(t \in d)|$ is the number of documents that feature the term t.

Suppose there is a threshold N, then we say that the term t characterizes a cluster c if:

$$P(d \in c | t \in d) > N * \max_{c_i \in C} \left(P(d \in c_i | t \in d) \right).$$

Thus, for every document d all terms that characterize the cluster of a document and are included in this document are seen as keywords K_d if the estimation $P(d \in c | t \in d)$ is high enough.

In order to extract L-tags we need to estimate the semantic function (13) value. The value of the semantic function (13) for a random document d and an L-tag lt is estimated as follows:

$$\sum_{k \in K_d \wedge k \in lt} P(d \in c | k \in d) \tag{26}$$

6 Clustering

Suppose we are given a data set of documents D on the data set of keywords K. The objective is to find out the best number of clusters for the data set of documents D to be used while conducting the clustering later on.

We will use the clustering method k-means++ [12] which enables us to divide a set of documents into clusters k in linear time.

The clustering quality factor with a parameter k is the value of Q_k, which is equal to the sum of roof-mean-square deviations of the centers of the obtained clusters for N iterations. Thus the smaller is Q_k, the more stable and profound is the clustering.

If k is too big or too small, it will affect the quality of the follow-up keyword extraction. That is why when choosing the parameter k it is necessary to set the up-value limit k_1 and down-value limit k_2.

The best value of k is somewhere in between of k_1 and k_2. You should chose that value of k which minimizes the value of Q_k for N iterations.

For example: Suppose $k_1 = 8$, $k_2 = 16$, $n = 10$. Then it will take 80 iterations to find the best clustering stability-wise, with the minimum value of Q_k.

7 Abstracting

After conducting a semantic search on book texts we face the problem of the search results representation. We can use the whole text of a book that meets the search query, an annotation of a book or an abstract that has been automatically complied beforehand. The idea to generate an abstract of a book on a user's demand seems to be more preferable.

Abstracting is a process of building up a summary (abstract) of a document. Abstracting is used for visual search results representation. Abstracts can be static and dynamic.

Static abstracts are used to represent summarized information about the whole document. A static abstract is a one-time generated abstract which does not depend on a user's search requirements.

Dynamic abstracts are generated at the moment when a user gives a query and feature summarized information about the relevant for the query parts of a text.

Abstracts can be generated with the use of a dominant-based algorithm [13]. Such approaches are quite common. The dominants may feature those extracted L-tags which are sentences.

8   The System Architecture

Figure 1 demonstrates the interaction scheme of the semantic search on library data. We can emphasize the following components:

- **The data uploading module** receives data from two sources. The metadata source supplies the module with bibliographic records. The metadata reaches the RDF store where a user can get it with the help of SPARQL queries. RDF store is Jena-based. The document source provides book texts which are saved in the file system.
- **Hierarchical clustering and keyword extraction module** launches clustering of all the book texts in the system. The clustering results are the ground for keywords extraction. Then for each cluster starts the process of clustering on a data set of keywords, which results in new keywords being extracted for each cluster. These two processes are performed alternately, until there is a text hierarchy and a keywords data set.
- **Indexes and Abstract generation module.** L-tags include sentences with keywords. According to the extracted keywords, L-tags get extracted and the semantic function (13) gets estimated. A set of the most valuable L-tags is formed for each document to be used later when generating its abstract. The extracted L-tags are indexed into Postgres DBMS. Each L-tag gets a list of documents linked to it.

To interact with the system a user should put a query in a natural language. Before being fulfilled, the query enriches itself with a lot of synonyms and hyponyms from the thesaurus. The L-tag search is conducted with the help of GIST and GIN. The similarity function (1) is estimated with the Okapi BM25 relevance-defining algorithm [14]. After the query the system looks for the relevant L-tags with the lists of linked documents. A dynamic abstract is formed for each document. This abstract represents an ordered set of the document L-tags which are relevant to the user's query.

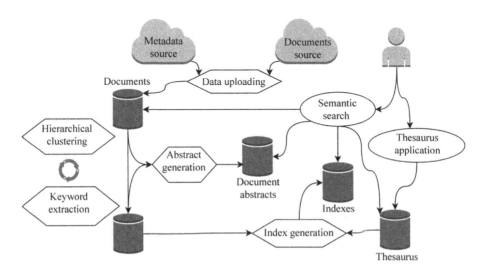

Fig. 1. Interaction scheme

A user can also search for documents with SPARQL queries in RDF store.

To improve the results of the search a user can edit the thesaurus that features synonyms and hyponyms.

9   Conclusion

The research featured the realization of a prototype semantic search system on bibliographic data and book texts.

Taking the semantic LibMeta library as an example we demonstrated the scientific relevance on the research. Implementing the proposed strategies helps to improve the library services quality.

There have been considered different ways of text semantic search realization. There has been implemented the L-tag model, the similarity function (1) and the semantic function (13). There has also been covered the semantic function interpolation (20). The query search objective has been reduced to the objective of similarity function (1) and semantic function (13) estimation. The proposed model was implemented when conducting search on the Postgres DBMS.

The work focused of the L-tag extracting algorithm which needs a data set of extracted keywords to operate.

There has been demonstrated a keyword hierarchy building process with the help of iterative alteration of clustering and keywords extraction. The proposed keywords extraction algorithm enables us to use the information on a cluster of a document. Clustering is performed with the k-means++ algorithm.

To visually represent the semantic search on texts results there has been demonstrated the approach to static and dynamic abstracting.

The proposed algorithms can be improved by the existent solutions, still they were deliberately simplified in the prototype framework. The further research will involve:

- Improving quality of the keywords extracting and the abstract generating algorithms.
- Experiments to improve the clustering quality.
- Experiments on L-tags extraction with the use of UDC hierarchy.
- Allocated keywords extraction and clustering on the Hadoop cluster.
- Distributed L-tag search system.

References

1. Ataeva, O.M., Serebryakov, V.A.: Personal digital LibMeta library as an open linked data integration environment. In: RCDL-2014 (2014). http://ceur-ws.org/Vol-1297/042-47_paper-8.pdf
2. Malakhov, D.A., Sidorenko, Y.A., Ataeva, O.M., Serebriakov, V.A.: Semantic search as a means of interaction with the digital library. In: DAMDID/RCDL, pp. 148–155 (2016). http://ceur-ws.org/Vol-1752/paper14.pdf

3. Giannopoulos, G., Bikakis, N., Dalamagas, T., Sellis, T.: GoNTogle: a tool for semantic annotation and search. In: Aroyo, L., Antoniou, G., Hyvönen, E., Teije, A., Stuckenschmidt, H., Cabral, L., Tudorache, T. (eds.) ESWC 2010. LNCS, vol. 6089, pp. 376–380. Springer, Heidelberg (2010). doi:10.1007/978-3-642-13489-0_27

4. Bontcheva, K., Tablan, V., Cunningham, H.: Semantic search over documents and ontologies. In: Ferro, N. (ed.) PROMISE 2013. LNCS, vol. 8173, pp. 31–53. Springer, Heidelberg (2014). doi:10.1007/978-3-642-54798-0_2

5. Berlanga, R., Nebot, V., Pérez, M.: Tailored semantic annotation for semantic search. Web Seman. Sci. Serv. Agents World Wide Web, 69–81 (2015). doi:http://dx.doi.org/10.1016/j.websem.2014.07.007

6. Alahmari, F., Magee, L.: Linked data and entity search: a brief history and some ways ahead. In: Proceedings of the 3rd Australasian Web Conference (2015). http://crpit.com/confpapers/CRPITV166Alahmari.pdf

7. Maynard, D., Greenwood, M.A.: Large scale semantic annotation, indexing and search at the national archives. In: LREC, pp. 3487–3494 (2012). https://gate.ac.uk/sale/lrec2012/tna/tna.pdf

8. Zakharova, I.V.: Speaking of a way to realize semantic document search on digital libraries. Ufa State Aviation Technical University Bulletin (2009). http://cyberleninka.ru/article/n/ob-odnom-podhode-k-realizatsii-semanticheskogo-poiska-dokumentov-v-elektronnyh-bibliotekah

9. Voskresensky, A.L., Khakhlin, G.K.: Semantic search tools. In: The 'Dialog' International Conference on Computer Linguistics and Intellectual Technologies (2006). http://www.nkj.ru/prtnews/29136/, http://www.dialog-21.ru/digests/dialog2006/materials/html/Voskresenskij.htm

10. Lukashevich, N.V.: Thesauruses in the information search problems (2010). http://www.nsu.ru/xmlui/handle/nsu/9086

11. Hasan, K.S., Ng, V.: Automatic keyphrase extraction: a survey of the state of the art. ACL **1**, 1262–1273 (2014). doi:10.3115/v1/P14-1119

12. Arthur, D., Vassilvitskii, S.: k-means++: The advantages of careful seeding. In: Proceedings of the Eighteenth Annual ACM-SIAM Symposium on Discrete Algorithms, pp. 1027–1035. Society for Industrial and Applied Mathematics (2007). doi:10.1145/1283383.1283494

13. Chanyshev, O.G.: Dominant association fields and text analysis. Sobolev Institute of Mathematics SB RAS (2011). doi:10.1145/1238844.1238847

14. Mayfield, J., McNamee, P.: Indexing using both n-grams and words. In: TREC, pp. 361–365 (1998). http://citeseerx.ist.psu.edu/viewdoc/summary?doi=10.1.1.61.899

Knowledge and Learning Management

Digital Ecosystem OntoMath: Mathematical Knowledge Analytics and Management

Alexander Elizarov[1(✉)], Alexander Kirillovich[1], Evgeny Lipachev[1], and Olga Nevzorova[2]

[1] Kazan (Volga Region) Federal University, 18 Kremlyovskaya ul.,
Kazan 420008, Russian Federation
amelizarov@gmail.com, alik.kirillovich@gmail.com,
elipachev@gmail.com

[2] Research Institute of Applied Semiotics of Tatarstan Academy of Sciences,
36a Levo-Bulachnaya ul., Kazan 420111, Russian Federation
onevzoro@gmail.com
http://kpfu.ru/
http://www.ips.antat.ru

Abstract. A mathematical knowledge management technology is discussed, its basic ideas, approaches and results are based on targeted ontologies in the field of mathematics. The solution forms the basis of the specialized digital ecosystem OntoMath which consists of a set of ontologies, text analytics tools and applications for managing mathematical knowledge. The studies are in line with the project aimed to create a World Digital Mathematical Library whose objective is to design a distributed system of interconnected repositories of digitized versions of mathematical documents.

Keywords: OntoMath · World digital mathematical library · WDML · Digital libraries · DML · Ontology · Linked data · Information retrieval · Mathematical content search · Semantic search · Formula search

1 Introduction

At the present time information and communication technologies are actively implemented in research and development. Therefore, it became possible to use the entire corpus of accumulated scientific knowledge in conducting new research. Such use requires creation of complex of technologies that ensure optimal management of available knowledge, the organization has effective access to this knowledge, as well as sharing and multiple use of new kinds of knowledge structures. In mathematics considerable experience in using of electronic mathematical content within various projects on creation of mathematical digital libraries is accumulated.

Since inception of the first scientific information systems, mathematicians have been involved in the full cycle of software product development, from idea to implementation. Well-known examples are an open source system TeX [1] and commercial systems Wolfram Mathematica and WolframAlpha, led by Stephen Wolfram according to his

L. Kalinichenko et al. (Eds.): DAMDID/RCDL 2016, CCIS 706, pp. 33–46, 2017.
DOI: 10.1007/978-3-319-57135-5_3

principles of computational knowledge theory [2, 3]. Tools for mathematical content management are developed with the help of communities of mathematicians, e.g. MathJax by American Mathematical Society, information system Math-Net.Ru is developed at the Steklov Mathematical Institute of the Russian Academy of Sciences [4] and the collection of publicly available preprints arXiv.org (https://arxiv.org/).

Main challenges of mathematical knowledge management (MKM) are discussed in [5–8]. In [9] we discuss the most urgent tasks: modeling representations of mathematical knowledge; presentation formats; authoring languages and tools; creating repositories of formalized mathematics, and mathematical digital libraries; mathematical search and retrieval; implementing math assistants, tutoring and assessment systems; developing collaboration tools for mathematics; creating new tools for detecting repurposing material, including plagiarism of others' work and self-plagiarism; creation of interactive documents; developing deduction systems. The solution of this task requires formalization of mathematical statements and proofs. While mathematics is full of formalisms, there is currently none of widely accepted formalisms for computer mathematics.

At the present time one of the largest formal mathematical libraries is Mizar (http://www.mizar.org/), which is a collection of papers prepared in the Mizar system of formal language), containing definitions, theorems and proofs [10, 11]. Mizar is one of the pioneering systems for mathematics formalization, which still has an active user community. The project has been in constant development since 1973.

Note the important results related to the level of formalization of representations of mathematical articles. For these purposes, developed languages of presentation of mathematical texts, specialized formal languages, as well as conversion software languages [12–16]. These technologies are also used to construct a mathematical ontology and creating semantic search service [6, 9, 17]. Effective communication requires a conceptualization as well as the sharable vocabulary. Ontologies that satisfy this requirement are described in [6, 9, 18].

In mathematical journals important part of the search service is to find fragments of formulas. For example, such a service is implemented in a digital repository Lobachevskii Journal of Mathematics (http://ljm.kpfu.ru/). For this we use converting documents in XML-format and formulas in of MathML-notation [12]. The above and many other mathematical implemented projects paved the way for the realization of a new idea, the creation of the World Digital Mathematical Library (WDML).

At the present time it formed a special type of information system called "digital ecosystem" [19]. The Digital Ecosystem is forming as the Information Technology, Telecommunications, and Media and Entertainment industries converge, users evolve from mere consumers to active participants, and governments face policy and regulatory challenges [20]. Research on digital ecosystems model adaptation to scientific and educational fields are described in [21–24].

This paper is devoted the development of the digital ecosystem OntoMath, whose task is the mathematical knowledge management in digital scientific collections. This ecosystem is a semantic publishing platform, which forms the semantic representation for the collections of mathematical articles and the set of ontologies and mathematical knowledge management services.

This article is an extension of the report "Mathematical Knowledge Management: Ontological Models and Digital Technology" at the conference "Data Analytics and Management in Data Intensive Domains" conference (DAMDID) [17]. We have expanded the description of the construction of Digital Mathematical Library technologies and we presented Object Paradigm of Mathematical Knowledge Representation. We have described the new tools a semantic search and knowledge management developed by us based on ontologies.

The paper is organized as follows. In Sect. 2, we consider the problems related to the management of digital mathematical libraries content. In Sect. 3 we present the object paradigm of representing mathematical knowledge. Then we present the architecture and the tools OntoMath ecosystem.

2 Digital Mathematics Libraries

At present, research activities in the field of mathematics associated with the use of modern information technology (cloud, semantic, etc.). These technologies are used in research of distributed scientific teams, the preparation and dissemination of mathematical knowledge in electronic form, the formation of mathematical digital libraries and of intellectual processing of their content. Special attention is given to the creation of a common information space by mathematical integration of existing and organizing new digital mathematical library (DML). The largest projects are "All-Russian Mathematical Portal Math-Net.RU" (http://www.mathnet.ru/), "Centre de diffusion de revues académiques mathématiques" (CEDRAM, http://www.cedram.org/), "Czech Digital Mathematics Library" (DML-CZ, http://dml.cz/), "The Polish Digital Mathematics Library" (DML-PL, http://pldml.icm.edu.pl/pldml/), "Göttinger DigitalisierungsZentrum" (GDZ, http://gdz.sub.uni-goettingen.de/gdz/), "Numérisation de documents anciens mathématiques" (NUMDAM, http://www.numdam.org/), Zentralblatt MATH (https://zbmath.org/), EMIS ELibM (http://www.emis.de/elibm/), "Bulgarian Digital Mathematics Library" (BulDML, http://sci-gems.math.bas.bg/jspui/) (see, e.g., [25–27]). Mathematical content is presented in a multidisciplinary digital libraries, for example, JSTOR (http://www.jstor.org/) and eLIBRARY (http://elibrary.ru/defaultx.asp).

This class also includes information of scientific publishing platform Elsevier (https://www.elsevier.com/), Springer (http://link.springer.com/), Pleiades Publishing (http://pleiades.online/ru/publishers/), as well as system support of scientific journals, for example, Elpub (http://elpub.ru/).

Realization and development of digital mathematical libraries involve the development of special tools and continuous improvement of their functionality. An example is the Open Journal Systems (OJS, https://pkp.sfu.ca/ojs/). The platform used in many projects, particularly in Lobachevskii Journal of Mathematics (http://ljm.kpfu.ru/), one of the first digital mathematical journals. In the practice of this journal, intelligent information processing tools since 1998, were introduced [12, 28–31]. In particular, we performed automated MathML-markup articles of this journal [12, 32]. Paper [33] presents a system of services for the automated processing of large collections of scientific documents. These services provide verification of document compliance to the accepted rules of formation of collections and their conversion to the established formats; structural analysis of documents and extraction of metadata, as well as their integration into the scientific information space. The system allows to automatically perform a set of operations that cannot be realized in acceptable time with the traditional manual processing of electronic content. It is designed for the large collections of scientific documents.

The idea of creating a World Digital Mathematical Library (WDML) arose in 2002. The initial aim of this project was digitizing the entire set of mathematical literature (both modern and historical), link it to the present literature, and make it clickable (see [25, 34–37]). As noted in [35], the success of this project and its future impact on mathematics, science and education could be the most significant event since the invention of scientific journals and to become a prototype for a new model of scientific and technical cooperation, a new paradigm for the future of science electronically connected world. At the same time, the implementation of such a large project will inevitably cause a series of problems. These problems and ways to overcome them were analyzed in [38]. In particular, one of the recommendations was the proposal to develop and coordinate some local projects of creating DML (see [26, 38]).

Basic plans for the construction of WDML in 2014–2015 discussed various mathematical communities and enshrined in a number of documents (see [39, 40]). In particular, it was noted that the next step in the development of the project WDML will be building information networks, knowledge-based, contained in mathematical publications. The discussion of these ideas was attended by many research groups of mathematicians all over the world, including our group of Kazan Federal University. In February 2016 in the Fields Institute (Toronto, Ontario) by the Wolfram Foundation, the Fields Institute, and the IMU/CEIC working group for the creation of a World Digital Mathematics Library it was organized the Semantic Representation of Mathematical Knowledge Workshop (https://www.fields.utoronto.ca/programs/scientific/15-16/semantic/). Our report on this symposium was devoted to the modeling and software solutions in the area of semantic representation of mathematical knowledge [41]. These results correspond to the general ideology WDML project of part semantic representation and processing of mathematical knowledge and are a strategic direction of research of our group. In particular, they are connected with the construction of OntoMath ecosystem, which is described below.

3 Moving Towards the Object Paradigm of Mathematical Knowledge Representation

E-libraries as a collection of electronic documents provide a document search by their bibliographic descriptions and thematic classification codes, as well as full-text search within the documents by keywords. Creating a full text index is a main mechanism for text search.

The global initiative WDML specifies key areas related to both organizational efforts of the international mathematical community, including mathematical literature publishers, research and technology efforts aimed at development of existing and intro-duction of new (semantic) technologies of representation and processing of mathemat-ical content. These semantic technologies include the following features:

- Aggregation of different ontologies, indexes, and other resources created by the mathematical community, and ensuring broad access to their replenishment and editing;
- Improving the access to mathematical publications – not only to searching and browsing, but for annotating, navigation, linking to other sources, data computing, data visualization, and so on.

The move towards the representation of the internal structure of mathematical knowl-edge creates a new paradigm of representation. The focus of representation has shifted to the selection of elements (classes) and their relationships, which allows researchers to create various network conceptual frameworks (e.g. the citation graph, the graph of mathematical concepts, etc.). Classification of mathematical objects and organization of the relevant repo-sitories provide new computing capabilities for data processing such as extraction and processing of formulas, finding similar papers and so on.

WDML project is focused on the object system of organization and storage of math-ematical knowledge. Unlike traditional electronic mathematical libraries in which the unit storage in the database is an electronic document, it is proposed to provide the mathematical knowledge of the collections of documents in the form of specially organ-ized repository of mathematical objects.

One of the key ideas is to develop the classes of objects for adequate description and study of mathematical content. In a mathematical document it is easy enough to identify a set of basic classes of mathematical objects (sequences, functions, transformations, identities, symbols, formulas, theorems, statements, etc.). As noted in WDML project, one of the most important tasks is to build a list of mathematical objects in different areas of mathematics.

Standard classes of mathematical objects are theorems, axioms, proofs, mathemat-ical definitions, etc. Important elements of the object model are semantic links (relations) between the elements. In order to build a document object model, it is proposed to use modern technologies of the Semantic Web. This representation of mathematical knowl-edge requires development of new management tools that will be relevant to mathe-matical knowledge (aggregation tools, semantic search, search formulas and identifica-tion of similar objects) [36, 39, 40].

Let us consider the key objects of mathematical knowledge management tools. Aggregation tools provide automatic collection of objects that meet certain criteria, as well as automatic replenishment of object lists. Object lists can be built according to different criteria, depending on the target application. For example, a useful list is a list of objects of a given domain (for example, a list of all known theorems of the group theory), or a list of objects of a particular class (e.g. theorem) related to the study of the mathematical properties of a given object (e.g. the geometric object "triangle"). These lists allow one to actually creating custom search indexes which would accumulate mathematical knowledge.

Navigation tools (with search tools) provide opportunities for navigation to target objects within the document. For example, the classical task is to find a given mathematical object and its properties, and to search for this given mathematical object and other mathematical objects related to certain mathematical equations. Another important task is to find a given mathematical object and scientific articles on this subject. At the same time, in contrast to the keyword search, object search would allow to take into account the semantics of links for object search, thereafter to improve search results.

For example, using the object properties for a given mathematical object (e.g., "Sobolev space") it is possible to find and view relevant information about such properties as its mathematical definition, educational literature, context-related objects and others.

Semantic search is the method of information retrieval which determines the relevance of the document to the query semantically rather than syntactically. Semantic search in the object repository is organized by following semantic links that allow to find objects by their description (implicit reference to object), as well as by given object properties. For example, the following query is classified as an implicit reference to the object: "Find all the theorems, the proof of which uses Fermat's theorem."

Search by formulas: this search tool provides search of mathematical formulas and additional information about them (such as the name of the formula, the list of scientific and educational publications, etc.). Formula search queries, in general, can have different forms. For example, a text query to the variables ("Find a formula connecting the area of the circle and the length of its circumference"), or computing request ("Find a formula equivalent to the formula, the F"), or text query to the mathematical object connected with this formula ("Find evidence of Euler's formula").

Identification tools are designed to identify identical objects that are referred to by different names and with different notations.

Thus, the main purpose of WDML is to unite digital versions of all mathematical repositories, including both contemporary sources and sources that have become historical, on new conceptual base and to provide intelligent information retrieval and data processing [39, 40].

At the same time new ways to detect objects of scientific knowledge directly through the web, as well as tools and services for creating and sharing of new types of knowledge structures are becoming more popular in the scientific community. In the context of the concept of Linked Data, and the Semantic Web these tools and services can be used to create "cooperation graphs" (collaboration graph), which are used, for example, to calculate the collaboration distance between the authors and searching similar documents. These facilities open up new possibilities of fine-tuning searching and browsing (see, e.g., [42]). Many authors (e.g. [6, 9, 18]) highlight the importance of developing new domain ontologies, in particular in mathematics, because the traditional bibliographic classification is no longer sufficient. It needs a deeper representation that would contain more detailed descriptions by taking into account different points of view.

4   OntoMath Digital Ecosystem

4.1   General Description

OntoMath is a digital ecosystem of ontologies, textual analytics tools, and applications for mathematical knowledge management. This system consists of the following components:

- Mocassin, an ontology of structural elements of mathematical scholarly papers;
- OntoMath[PRO], an ontology of mathematical knowledge concepts;
- Semantic publishing platform;
- Semantic formula search service;
- Recommender system.

Briefly we describe these basic elements of the architecture of digital ecosystems OntoMath (Fig. 1).

Fig. 1. OntoMath ecosystem architecture

The core component of the OntoMath ecosystem is its semantic publishing platform. It builds an LOD representation for a collection of mathematical articles in LaTeX. The generated mathematical dataset includes metadata, the logical structure of documents, terminology, and mathematical formulas. Article metadata, the logical structure of documents, and terminology are expressed in terms of AKT Portal, Mocassin and OntoMath[PRO] ontologies, respectively. Mocassin ontology, in its turn, is built on SALT Document Ontology, which is ontology of the rhetorical structure of scholarly publications. Mocassin and OntoMath[PRO] ontologies are parts of OntoMath ecosystem but SALT is an external ontology. Two applications are built using the semantic publishing platform: a semantic formula search service and a recommender system.

As any digital ecosystem, OntoMath has components that are used for sociotechnical purposes. Such components are ontologies and semantic publishing platforms. They can be used by mathematicians and software systems developers.

4.2   Semantic Publishing Platform

As was mentioned above, a semantic publishing platform, which constitutes the core of the OntoMath ecosystem makes an LOD representation for a given sample of mathematical articles in LaTeX [43, 44]. Its main features are:

- Indexing mathematical articles in LaTeX-format as LOD-compatible RDF-data;
- Extracting articles' metadata in terms of AKT Portal Ontology;
- Mining the document logical structure using our ontology of structural elements of mathematical papers;
- Eliciting instances of mathematical entities as the concepts of OntoMath^{PRO} ontology;
- Connecting the extracted textual instances to symbolic expressions and formulas in the mathematical notation;
- Establishing the relationship between published data and RDF-existing sets of LOD data.

The developed technology has the following features:

- Mathematics RDF-set is based on a collection of mathematical articles in Russian;
- The RDF-built set that includes metadata and also specific semantic knowledge such as the knowledge generated as a result of special treatment of mathematical formulas (binding textual definitions of variables with the symbols of variables in formulas) and also the instances of OntoMathPRO ontology and the structural elements of the mathematical articles.
- Semantic annotation of mathematical texts based on Mocassin and OntoMath^{PRO} ontologies.
- The MathLang Document Rhetorical (DRa) Ontology [45] enables one to interpret the elements of document structure using mathematical rhetorical roles that are similar to the ones defined in the statement level of OMDoc ontology. This semantics focuses on formalizing proof skeletons for generation of proof checker templates.

4.3   Ontologies

Mocassin (https://code.google.com/archive/p/mocassin/) is an ontology intended to annotate a logical structure of a mathematical document [43, 46]. This ontology extends SALT Document Ontology, defining concepts and relations specific to mathematical documents. Mocassin ontology represents a mathematical document as a set of interconnecting segments. It is designed using OWL2/RDFS [47] languages, which provided with expressive possibilities, as well as theoretical and practical output means.

Ontology Mocassin uses 15 concepts such as *Document segment, Claim, Definition, Proposition, Example, Axiom, Theorem, Lemma, Proof, Equation*, and others. The ontology defines relations between segments such as *dependsOn, exemplifies, hasConsequence, hasSegment, proves, refersTo*.

OntoMatn^{PRO} (http://ontomathpro.org/) is an ontology of mathematical knowledge [9, 48, 49]. Its concepts are organized into two taxonomies:

- Hierarchy of areas of mathematics: *Logics*, *Set theory*, *Geometry*, including its sub-fields, such as *Differential Geometry,* and so on;
- Hierarchy of mathematical objects such as a *set, function, integral, elementary event, Lagrange polynomial,* etc.

This ontology defines the following relations:

- Taxonomic relation (for example, "Lambda matrix" *is a* "Matrix");
- Logical dependency (for example, "Christoffel Symbol" *is defined by* "Connectedness");
- Associative relation between objects (for example, "Chebyshev Iterative Method" see also "Numerical Solution of Linear Equation Systems");
- Belongingness of objects to fields of mathematics (for example, "Barycentric Coordinates" *belongs to* "Metric Geometry");
- Associative relation between problems and methods (for example, "System of linear equations" *is solved by* "Gaussian elimination method").

Each concept description has Russian and English labels, textual definitions, and relations with other concepts, links to external terminologies, such as DBpedia and ScienceWISE.

4.4 Applications

OntoMath Formula Search Engine is a semantic search service that uses a semantic representation of math document built on the base of Semantic Publishing Platform [48]. OntoMath Formula Search Engine implements new search on names of variables using OntoMath ontology. A variable in the formula is a symbol that denotes a mathematical object. Mathematical symbols can denote numbers (constants), variables, operations, functions, punctuation, grouping, and other aspects of logical syntax. Specific branches and applications of mathematics usually have specific naming conventions for variables. However, nonstandard names of variables may be used in some formulas. OntoMath Formula Search Engine allows finding the mathematical formulas containing a given mathematical object regardless of its name for the variable. For example, if we would like to find a formula that contains a mathematical object (e.g. the curvature), the service will find all the formulas that include this object (even with different names for the variable). Using an inference, the service can find the formulas containing not only the given object, but the objects below in the hierarchy of the ontology. For example, for searching the formulas which contain the polygon, OntoMath Formula Search can find the formulas which contain not only the polygon but other objects in the hierarchy (e.g. the triangle, the parallelogram, the trapezium, the hexagon and others). OntoMath formula search also allows restricting your search to the document area that you define. For example, you can search only in the defined areas or in a certain theorem area. These search functions of OntoMath Formula Search Engine differ from those of popular search services, such as (uni) quation, Springer LaTeX search, Wikipedia search formula, Wolfram search formula. These services have a great potential, including their stability for renaming variables and for expression transformation. However, they are syntactic and seek formulas containing a predetermined formula pattern.

We have implemented two applications for mathematical formula search such as syntactical search of formulas in MathML, and semantic ontology-based search.

The syntactical search leverages formula description from documents formatted in TeX. Our algorithm [12] transforms formulas from TeX format to MathML format. We set up an information retrieval system prototype for a collection of articles in Loba-chevskii Journal of Mathematics. For the end-user, the query input interface supports a convenient syntax. The search results include highlighted occurrences of formulas as well as document metadata.

OntoMath Recommender System

As ecosystem OntoMath application we have developed a recommender system for the collection of physical and mathematical documents. One of the main functions of this system is the creation of the list of related documents (see [50]). Traditionally, the list of related documents is based on the keywords given by the authors, as well as biblio-graphic references available in the documents.

This approach has several disadvantages:

- A list of keywords may be missing or incomplete;
- A keyword may be ambiguous;
- It is necessary to take into account the hierarchy of concepts.
- It should be noted that the article may use the terminology in different languages.

Thus, the created recommender system has the following features:

- It takes into account the professional profile of a particular user;
- It forms different recommendations for different scenarios of work with the system (referee, user being introduced in the topic, etc.);
- It assigns different weights to different concepts. Thus, for a scientific review, concepts denoting areas of mathematics are more important than those related to mathematical objects. At the same time, for a beginning researcher survey papers containing notions from different areas of mathematics and references to original works are more important.

Recommender system's workflow consists of the following steps:

- Ontology-based keywords extraction;
- Semantic representation of an electronic collection of mathematical papers;
- Calculation of the measure of thematic proximity between documents by using this representation;
- Building a list of recommended papers.

5   Conclusion

The basic ideas, approaches and results of developing the discussed mathematical knowledge management technology are based on targeted ontologies in the field of mathematics. These solutions form the basis of the specialized digital ecosystem Onto-Math which consists of a set of ontologies, text analytics tools and applications for managing mathematical knowledge. The studies are in line with the project aimed to create a World Digital Mathematical Library whose objective is to design a distributed system of interconnected repositories of digitized versions of mathematical documents.

The future of the OntoMath ecosystem is related to the development of new services for semantic text analytics and control of mathematical knowledge. The developed technologies are supposed to be evaluated with the help of the digital mathematical collections of Kazan Federal University.

The present work is aimed at further research in the field of mathematical knowledge management; it has been carried out by the authors of the paper since 1998, with the support of grants from the Russian Foundation for Basic Research, Kazan Federal University and the Academy of Sciences of the Tatarstan Republic.

Acknowledgement. This work was funded by the subsidy allocated to Kazan Federal University for the state assignment in the sphere of scientific activities, grant agreement no. 1.2368.2017).

References

1. Knuth, D.E.: The TeX Book. Addison-Wesley Publishing Company, Reading (1986)
2. Wolfram, S.: A New Kind of Science. Wolfram Media Inc., Champaign (2002)
3. Wolfram, S.: An Elementary Introduction to the Wolfram Language. Wolfram Media Inc., Champaign (2015)
4. Chebukov, D.E., Izaak, A.D., Misyurina, O.G., Pupyrev, Y.A., Zhizhchenko, A.B.: Math-Net.Ru as a digital archive of the russian mathematical knowledge from the XIX century to today. In: Carette, J., Aspinall, D., Lange, C., Sojka, P., Windsteiger, W. (eds.) CICM 2013. LNCS (LNAI), vol. 7961, pp. 344–348. Springer, Heidelberg (2013). doi: 10.1007/978-3-642-39320-4_26
5. Carette, J., Farmer, William M.: A review of mathematical knowledge management. In: Carette, J., Dixon, L., Coen, C.S., Watt, Stephen M. (eds.) CICM 2009. LNCS (LNAI), vol. 5625, pp. 233–246. Springer, Heidelberg (2009). doi:10.1007/978-3-642-02614-0_21
6. Lange, C.: Ontologies and languages for representing mathematical knowledge on the semantic web. Semant. Web 4(2), 119–158 (2013). doi:10.3233/SW-2012-0059
7. Ion, P.D.F.: Mathematics and the world wide web. In: Carette, J., Aspinall, D., Lange, C., Sojka, P., Windsteiger, W. (eds.) CICM 2013. LNCS (LNAI), vol. 7961, pp. 230–245. Springer, Heidelberg (2013). doi:10.1007/978-3-642-39320-4_15
8. Kohlhase, M.: Mathematical knowledge management: transcending the one-brain-barrier with theory graphs. Newslett. Eur. Math. Soc. **92**, 22–27 (2014)
9. Elizarov, A., Kirillovich, A., Lipachev, E., Nevzorova, O., Solovyev, V., Zhiltsov, N.: Mathematical knowledge representation: semantic models and formalisms. Lobachevskii J. Math. **35**(4), 347–353 (2014). doi:10.1134/S1995080214040143
10. Naumowicz, A., Korniłowicz, A.: A brief overview of Mizar. In: Berghofer, S., Nipkow, T., Urban, C., Wenzel, M. (eds.) TPHOLs 2009. LNCS, vol. 5674, pp. 67–72. Springer, Heidelberg (2009). doi:10.1007/978-3-642-03359-9_5

11. Bancerek, G., Byliński, C., Grabowski, A., Korniłowicz, A., Matuszewski, R., Naumowicz, A., Pąk, K., Urban, J.: Mizar: state-of-the-art and beyond. In: Kerber, M., Carette, J., Kaliszyk, C., Rabe, F., Sorge, V. (eds.) CICM 2015. LNCS (LNAI), vol. 9150, pp. 261–279. Springer, Cham (2015). doi:10.1007/978-3-319-20615-8_17

12. Elizarov, A.M., Lipachev, E.K., Malakhaltsev, M.A.: Web technologies for mathematicians: the basics of MathML. In: A Practical Guide. Fizmatlit, Moscow (2010). (in Russian)

13. Kohlhase, M.: An Open Markup Format for Mathematical Documents (Version 1.2). LNAI, vol. 4180. Springer, Heidelberg (2006). http://omdoc.org/pubs/omdoc1.2.pdf

14. Iancu, M., Kohlhase, M., Rabe, F., Urban, J.: The mizar mathematical library in OMDoc: translation and applications. J. Autom. Reason. 50(2), 191–202 (2013). Springer

15. Kohlhase, M.: Semantic Markup in TeX/LaTeX. http://ctan.altspu.ru/macros/latex/contrib/stex/sty/stex/stex.pdf

16. Dehaye, P.-O., et al.: Interoperability in the OpenDreamKit project: the math-in-the-middle approach. In: Kohlhase, M., Johansson, M., Miller, B., de Moura, L., Tompa, F. (eds.) CICM 2016. LNCS (LNAI), vol. 9791, pp. 117–131. Springer, Cham (2016). doi: 10.1007/978-3-319-42547-4_9

17. Elizarov, A.M., Kirilovich, A.V., Lipachev, E.K., Nevzorova, O.A.: Mathematical knowledge management: ontological models and digital technology. In: CEUR Workshop Proceedings, vol. 1752, pp. 44–50 (2016). http://ceur-ws.org/Vol-1752/paper08.pdf

18. Staab, S., Studer, R. (eds.): Handbook on Ontologies. Springer, Heidelberg (2003). (Republished in 2009)

19. Briscoe, G., De Wilde, P.: Digital ecosystems: evolving service-oriented architectures. In: Conference on Bio Inspired Models of Network, Information and Computing Systems. IEEE Press (2006). arXiv:0712.4102v6. https://arxiv.org/pdf/0712.4102v6.pdf

20. Digital Ecosystem Convergence between IT, Telecoms, Media and Entertainment: Scenarios to 2015. http://www3.weforum.org/docs/WEF_DigitalEcosystem_Scenario2015_ExecutiveSummary_2010.pdf

21. Amritesh, C.J.: Digital ecosystem for knowledge, learning and exchange: exploring socio-technical concepts and adoption. In: Basile Colugnati, F.A., et al. (eds.) OPAALS 2010. LNICST, vol. 67, pp. 44–61. Springer, Heidelberg (2010)

22. David, C., Ginev, D., Kohlhase, M., Corn, J.: eMath 3.0: building blocks for a social and semantic Web for online mathematics & elearning. In: Mierluş-Mazilu I. (ed.) Proceedings of the 1st International Workshop on Mathematics and ICT: Education, Research and Applications, pp. 13–23. Conspress, Bucureşti (2010). http://kwarc.info/kohlhase/papers/malog10.pdf

23. David, C., Ginev, D., Kohlhase, M., Matican, B., Mirea, S.: A framework for semantic publishing of modular content objects. In: CEUR Workshop Proceedings, vol. 721, pp. 1–12 (2011). http://ceur-ws.org/Vol-721/paper-03.pdf

24. Kohlhase, M., Corneli, J., David, C., Ginev, D., Jucovschi, C., Kohlhase, A., Lange, C., Matican, B., Mirea, S., Zholudev, V.: The planetary system: web 3.0 & active documents for STEM. In: Sato, M., Matsuoka, S., Sloot, P.M., Dick Albada, G., Dongarra, J. (eds.) Procedia Computer Science (4) (Special issue: Proceedings of the International Conference on Computational Science), pp. 598–607. Elsevier (2011)

25. Bouche, T.: Digital mathematics libraries: the good, the bad, the ugly. Math. Comput. Sci. 3(3), 227–241 (2010). doi:10.1007/s11786-010-0029-2

26. Sylwestrzak, W., Borbinha, J., Bouche, T., Nowinski, A., Sojka P.: EuDML – towards the European digital mathematics library. In: Sojka P. (ed.) Towards a Digital Mathematics Library. Paris, 7–8 July 2010, pp. 11–26. Masaryk University Press, Brno (2010). http://dml.cz/bitstream/handle/10338.dmlcz/702569/DML_003-2010-1_5.pdf

27. Bouche, T.: Towards a World Digital Library: Mathdoc, Numdam and EuDML Experiences. UMI, La Sapienza, Roma (2016). http://www.mat.uniroma1.it/sites/default/import-files/biblioteca/SEMINARIO2016/bouche.pdf

28. Elizarov, A., Lipachev, E., Zuev, D.: Mathematical content semantic markup methods and open scientific e-journals management systems. In: Klinov, P., Mouromtsev, D. (eds.) KESW 2014. CCIS, vol. 468, pp. 242–251. Springer, Cham (2014). doi:10.1007/978-3-319-11716-4_22

29. Elizarov, A.M., Zuev, D.S., Lipachev E.K.: Open scientific e-journals management systems and digital libraries technology. In: CEUR Workshop Proceedings, vol. 1108, pp. 102–111 (2013). http://ceur-ws.org/Vol-1108/paper13.pdf

30. Elizarov, A.M., Zuev, D.S., Lipachev, E.K.: Infrastructure of electronic scientific journal and cloud services supporting lifecycle of electronic publications. In: CEUR Workshop Proceedings, vol. 1297, pp. 156–159 (2014). http://ceur-ws.org/Vol-1297/156-159_paper-23.pdf

31. Elizarov, A.M., Zuev, D.S., Lipachev, E.K.: Electronic scientific journal management systems. Sci. Tech. Inf. Process. 41(1), 66–72 (2014). doi:10.3103/S0147688214010109

32. Elizarov, A.M., Zuev, D.S., Lipachev, E.K., Malakhaltsev, M.A.: Services structuring mathematical content and integration of digital mathematical collections at scientific information space. In: CEUR Workshop Proceedings, vol. 934, pp. 309–312 (2012). http://ceur-ws.org/Vol-934/paper47.pdf

33. Elizarov, A.M., Lipachev, E.K., Haidarov, S.M.: Automated processing service system of large collections of scientific documents. In: CEUR Workshop Proceedings, vol. 1752, pp. 58–64 (2016). http://ceur-ws.org/Vol-1752/paper10.pdf

34. Jackson, A.: The digital mathematics library. Not. AMS 50(4), 918–923 (2003). http://www.ams.org/notices/200308/comm-jackson.pdf

35. The Digital Mathematical Library Project. Status, August 2005. http://www.math.uiucedu/~tondeur/DML04.pdf

36. Digital Mathematics Library: a Vision for the Future. International Mathematical Union (2006). http://www.mathunion.org/fileadmin/IMU/Report/dml_vision.pdf

37. Tondeur, P.: WDML: the world digital mathematics library. The evolution of mathematical communication in the age of digital libraries. In: IMA Workshop, December 8–9 (2006). http://www.math.uiuc.edu/~tondeur/WDML_IMA_DEC2006.pdf

38. Pitman, J., Lynch, C.: Planning a 21st century global library for mathematics research. Not. AMS 61(7), 776–777 (2014). http://www.ams.org/notices/201407/rnoti-p776.pdf

39. Developing a 21st century global library for mathematics research. Washington, D.C.: The National Academies Press, Washington, D.C (2014). arxiv.org/pdf/1404.1905, http://www.nap.edu/catalog/18619/developing-a-21st-century-global-library-for-mathematics-research

40. Olver, P.J.: The world digital mathematics library: report of a panel discussion. In: Proceedings of the International Congress of Mathematicians, 13–21 August 2014, Seoul, Korea, Kyung Moon SA, vol. 1, pp. 773–785 (2014)

41. Elizarov, A.M., Zhiltsov, N.G., Kirillovich, A.V., Lipachev, E.K., Nevzorova, O.A., Solovyev, V.D.: The OntoMath ecosystem: ontologies and applications for math knowledge management. In: Semantic Representation of Mathematical Knowledge Workshop, 5 February 2016. http://www.fields.utoronto.ca/video-archive/2016/02/2053–14698

42. Todeschini, R., Baccini, A.: Handbook of Bibliometric Indicators: Quantitative Tools for Studying and Evaluating Research. Wiley-VCH Verlag (2016)

43. Nevzorova, O.A., Birialtsev, E.V., Zhiltsov, N.G.: Mathematical text collections: annotation and application for search tasks. Sci. Tech. Inf. Process. 40(6), 386–395 (2013)

44. Nevzorova, O., Zhiltsov, N., Zaikin, D., Zhibrik, O., Kirillovich, A., Nevzorov, V., Birialtsev, E.: Bringing math to LOD: a semantic publishing platform prototype for scientific collections in mathematics. In: Alani, H., et al. (eds.) ISWC 2013. LNCS, vol. 8218, pp. 379–394. Springer, Heidelberg (2013). doi:10.1007/978-3-642-41335-3_24
45. Kamareddine, F., Wells, J.B.: Computerizing mathematical text with mathlang. Electron. Notes Theor. Comput. Sci. **205**, 5–30 (2008)
46. Solovyev, V., Zhiltsov, N.: Logical structure analysis of scientific publications in mathematics. In: Akerkar, R. (ed.) Proceedings of the International Conference on Web Intelligence, Mining and Semantics (WIMS 2011), vol. 21, pp. 1–9. ACM DL (2011). doi: 10.1145/1988688.1988713
47. OWL 2 Web Ontology Language. RDF-Based Semantics (Second Edition). W3C Recommendation 11 December 2012. https://www.w3.org/2012/pdf/REC-owl2-rdf-based-semantics-20121211.pdf
48. Nevzorova, O., Zhiltsov, N., Kirillovich, A., Lipachev, E.: OntoMath^{PRO} ontology: a linked data hub for mathematics. In: Klinov, P., Mouromstev, D. (eds) KESW 2014. CCIS, vol. 468, pp. 105–119. Springer, Heidelberg (2014). doi:10.1007/978-3-319-11716-4_9
49. Elizarov, A.M., Lipachev, E.K., Nevzorova, O.A., Solov'ev, V.D.: Methods and means for semantic structuring of electronic mathematical documents. Doklady Math. **90**(1), 521–524 (2014). doi:10.1134/S1064562414050275
50. Elizarov, A.M., Kirillovich, A.V., Lipachev, E.K., Zhizhchenko, A.B., Zhil'tsov, N.G.: Mathematical knowledge ontologies and recommender systems for collections of documents in physics and mathematics. Doklady Math. **93**(2), 231–233 (2016). doi:10.1134/S1064562416020174

Development of Fuzzy Cognitive Map for Optimizing E-learning Course

Vasiliy S. Kireev[✉]

National Research Nuclear University MEPhI (Moscow Engineering and Physics Institute),
Moscow, Russian Federation
vskireev@mephi.ru

Abstract. Learning management system (LMS) optimization has been one of the core issues in the face of increasing learning content supply and the rising number of online-course participants. This optimization is mostly based on LMS logs data analysis and revealing the users' behavior patterns linked to the content. This article focuses on the approach to LMS users' behavior pattern simulation that stems from the fuzzy cognitive maps-featuring approach. The proposed model describes the user-content interaction within the system and can be applied to predict users' reactions to its learning, testing and practical elements. The obtained cognitive map has been tested with the INFOMEPHIST system data. This system has been used to assist leaning process in a number of National Research Nuclear University MEPhI departments for more than nine years. The current and further research is supported by the NRNU MEPhI development program.

Keywords: Fuzzy cognitive maps · Moodle-based LMS · Users' Behavior Simulation · E-learning

1 Introduction

As of today e-learning has been gaining increasing attention both in the corporate sector (with the Corporate University concept being massively introduced) and among classic educational institutions [4]. Some MOOC platforms such as Coursera, EdX etc. should surely be noted as well. This pattern helps to boost learning process and to lower organization and production costs, as well as to make knowledge transmission automatic and to obtain additional sources of the information on education quality and students' behavior. There exist corporate and open source LMS. Among the latter, Moodle is the most popular. It is implemented in many universities as a framework for their own software. Moodle LMS is an object-oriented module-consisting dynamic learning platform. The system records student behavior data to a file and stores it in the database. This data is represented as a table that features the information on Course, Time, Student's Name, Action and Section. The latest years have seen e-learning accumulating more data which makes it relevant for the E-learning data mining. This is why boosting education process with the help of learning content optimization seems essential. This information can be used for making up a sufficient and detailed student behavior model.

L. Kalinichenko et al. (Eds.): DAMDID/RCDL 2016, CCIS 706, pp. 47–56, 2017.
DOI: 10.1007/978-3-319-57135-5_4

The data on his behavior can reveal his behavior pattern, which will then enable us to classify it (to relate it to a previously obtained student behavior pattern). Using this classification we can build up recommendations concerning course-related materials and boost the academic performance level on the course.

This work is an extension of [13]. Comparing with the previous work in this paper a modification of developed cognitive map is introduced. Additional simulations results with different initial conditions are presented and discussed.

2 Modern Views on the Problem

Most of the approaches to web-content and LMS optimization lie in revealing users' behavior patterns after analyzing the way they use the system (Web Mining) [2, 3]. The current analysis in mostly conducted through clustering and classification. Besides we use association rules mining, sequential and content analyses (see Fig. 1).

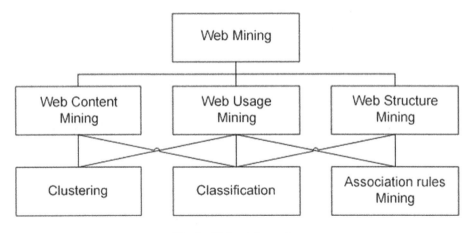

Fig. 1. Web mining today

The latest years have also seen cognitive visualization development [10], which enables us to describe users' studying trends with the LMS logs. All the methods listed above help to reveal local characteristics of users' behavior, although it is still quite rare when we can apply the obtained results to the whole user-LMS interaction process [5]. Thus, revealing the weak points of the system content calls for additional (mostly time-consuming) research.

Due to various reasons a course author does not always has the opportunity to conduct such a research, and the existing data remains unclaimed. Thus, meeting fundamental content requirements is also essential. These requirements should be laid down at the first stage of a course designing process to result into better quality content. To reach this goal we embark on making a generalized model that will allow us to define the optimal-quantity and optimal-quality forms of the educational content.

One of the new Decision Support System theory trend is cognitive modeling with semi-structured data and semi-structured cases management analysis. Cognitive structuring includes defining prospective targets of the object in control, as well as undesirable conditions and valuable environment factors that affect the transition of the object to its current state. It also includes revealing cause-effect relations, taking into account the mutual influence factor.

3   Fuzzy Cognitive Maps

Cognitive maps are one of the tools to represent ideas in weakly-formalized fields such as economics, politics and the military sphere. This approach was introduced in 1976 by Robert Axelrod in his book, dedicated to cognitive maps in politics [7]. A cognitive map is a symbolic directed graph that features vertices that represent entities, concepts, factors, targets and events, and arcs that show the influence of one vertex to another. A concept of a situation suggests that this situation develops in time and under changing environment that is reflected in changing factor values. A concept map enables us to set a problem of prediction (which is a direct task) and to explore possible ways to manage the situation, i.e. to find out what effects lead the desired (targeted) condition (which is an inverse task). The effect is characterized by a threshold function which can be defined in a variety of ways. The function includes an expert valuation, which is initially set in a natural language. Bart Kosko extended the paradigm by introducing fuzziness [1]. This reflects the dispersion of the experts opinions on the mutual influence of different factors. Fuzzy numbers are usually represented by triangular numbers.

On the whole the objective to define the vertices (concepts) state can be reduced to the following calculations (see Formula 1):

$$A_i(k + 1) = f\left(A_i(k) + \sum_{j \neq i, j=1}^{N} A_j(k)W_{ji} \right) \tag{1}$$

The calculating process is iterative, which means that after you set the initial vertices states their values are recalculated until the difference between the current states and the previous states is less that an ε value.

As of today, managing complex systems often implies using differently-formalized cognitive maps [6] on different stages of decision-making in semi-structured problem fields, especially in the social and economic spheres. Cognitive maps serve as the basis for map generation and map verification methods, which support the formation of a situation common knowledge bank. At this stage designing cognitive maps aims at visual representation of a problem [11, 12] to help explain a subject's actions, with his point of view analysis serving as the starting point. In this case whether the map is adequate or not can be confirmed by the subject. Different interpretations of edges, weights, vertices and various functions that define lines-factors influence make up diverse cognitive map modifications and ways to examine them.

The interpretations can differ both contently and mathematically. Due to the fact that cognitive maps have a great number of modifications, we can consider different types of models based on these maps. One of the modern trends in cognitive map development is frame-based cognitive maps [9].

4   The Proposed Approach

We suggest a fuzzy cognitive map as a user-LMS interaction model. It describes the way a set of concepts that characterize the content from the didactic and systematic point of view influence the student academic efficiency (Course Competence) [13]. The course in particular represents a set of modules (Module Competence), the acquisition level of which affects the Course Competence concept. A separate module includes a set of static and interactive content. Static learning content (Learning Competence) includes abstracts and lecture-related presentations, additional teaching materials, etc. Interactive content includes a solely controlling component of tests (Tests Competence) and a practically-oriented learning component (Practice Competence) of lab assignments and simulators that students run via the system (e.g. SCORM packages). Generally speaking, this kind of content depends on accomplishment time: it could formally be the same as Moodle Task component, but we include it into Practice only when the maximum accomplishment time does not exceed the standard 4-hours learning maximum.

Other entities include: the user-system interaction (LMS Interaction), the number of the user's log-ins (Entry Attempts), the Time, spent in the system (LMS Time), the Feedback Quality, the Number of new topics created by the user (Topics Number), the Number of messages sent by the user (Messages Number), the Results of other students (Listeners Results), the Final grade for the course (Course Mark), the Trajectory, which stands for the time the user accesses the modules and the content within them, corresponding to the natural sequence of learning course (FOR), the Recommended Sources, the Learning Materials, the Number of tests, the Number of attempts for passing a test (KP), the Time spent for the tests (Tests Time), etc. Initially proposed cognitive map was modified. Thus, there have been added 4 new concepts to obtain a more balanced result. The new map was created with the help of an open source Mental Modeller application [14] (see Fig. 2).

These concepts to some extent oppose the already existing elements, such as Exercises Number, Tests Number, Tasks Number, Materials Number, i.e. the High Tests Number concept positively influence the Tests Competence, but if number of tests is small or equals zero, it negatively tells on the Competence. To represent such a case there has been introduced a new Low Tests Number concept that has the negative weight connection with the Tests Competence. As far as a low and a high number cannot exist at the same time, these concepts were connected by the negative feedback edges (See Fig. 3).

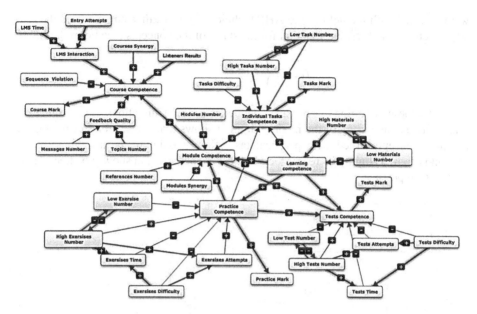

Fig. 2. A user-LMS interaction cognitive map (created with Mental Modeller)

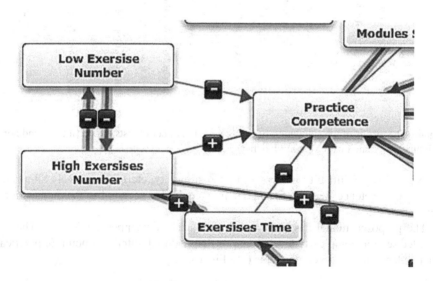

Fig. 3. The added concept

After consulting 5 e-learning course making experts, we have defined the weights of the arcs. We used simple linguistic 5-point scales, where 1 point stands for the lowest influence level and 5 points stand for the highest influence level. The experts grades were adjusted with the help of the AHP technique and fuzzificated. We used the unit step function as a threshold function for the states of the concepts (see Formula 2):

$$f(x) = \begin{cases} 1, x \geq 0 \\ 0, x \leq 0 \end{cases} \qquad (2)$$

Under these conditions were have conducted several simulations to determine the sensitivity and adequacy of the model. Figure 4 shows that if the values of the highly practical concepts, filled with practical tasks, home assignments and other related content is high, it slightly increases the Module Competence, but shows no effect on the Course Competence.

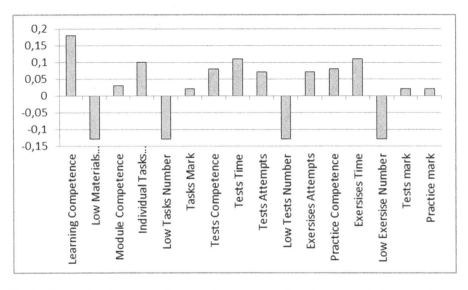

Fig. 4. The results of the simulation featuring a great number of tests, practical tasks and home assignments (created with Mental Modeller)

When diverse difficulty is increased (tests difficulty, etc.), it negatively affects the Module Competence (see Fig. 5) and the time required to accomplish these educational elements.

The proposed model has been tested with a free Fcmapper application. The test included several complexity level scenarios that featured different amount of practical, independent and testing components (see Fig. 6).

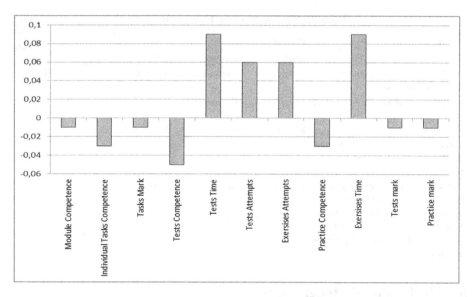

Fig. 5. The results of the simulation featuring high complexity level of of tests, practical tasks and home assignments (created with Mental Modeller)

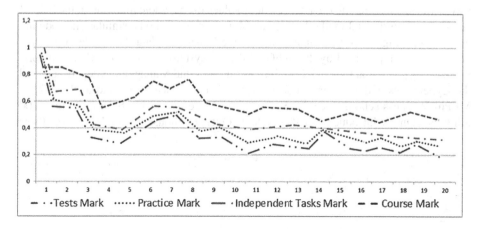

Fig. 6. The results of the low complexity level scenario

The line graph shows the main concepts, including the Course Competence, and we can see that the low complexity level lets the Course Competence value reach the acceptable number of 0,6. The display indicators (or course components marks) also fit within the normative values, including those of the ECTS system.

Table 1 represents a part of the simulation results, which show that increasing learning elements number and their difficulty does not lead to increasing Course Competence.

Table 1. The results of the FCM simulation

Factor	Competence			
	Tests	Practice	Independent Tasks	Course
The Number of tests, exercises and tasks is the Highest	Increases	Increases	Increases	No effect
The Difficulty of tests, exercises and tasks is the Highest	Decreases	Decreases	Decreases	No effect
The Modules Number, the Modules Synergy, the Materials Number and the References Number are the Highest	Increases	Increases	Increases	Increases
The Number of tests, exercises and tasks, the Difficulty of tests and exercises, the Modules Number, the Modules Synergy, the Materials Number and the References Number are the Highest	Increases	Increases	Increases	Increases

5   The Proposed Model Testing

Cognitive map verification is a nontrivial task which is usually solved in two different ways [8]: (1) checking the obtained vertices values or the whole model with the help of alternative patterns and methods, such as Monte Carlo methods, simulation modeling etc.; (2) checking each edge conclusion on the real historical data. The model was tested on the real data, provided by the INFOMEPHIST system (see Fig. 7), which supports the learning process at the NRNU MEPhI Economics and Management department and the Cybernetics and Information security department since 2007, and at the NRNU MEPhI Business school since 2015. During this time there have been more than 100 different courses introduced in the system. Over 15 thousands students took these

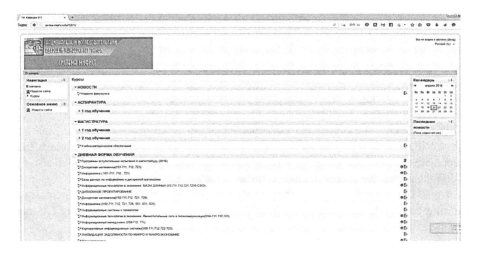

Fig. 7. The INFOMEPHIST home page

courses. The INFOMEPHIST system is based on the open source Moodle LMS, which enables us to manage the content and the users, and to monitor their activity. Thus, there has been accumulated much information about users behavior that was later analyzed to make the proposed cognitive map parameters more clear. The system register includes more than 3 million records.

The analysis used the system data corresponding to the curriculum of the master's program in Economics. With the help of correlation and regression approach there have been estimated several proposed entities, such as the indicator elements (practical component marks and other marks) against the second order entities, such as the SCORM laboratory works number (EN), the number of attempts taken (AN), the total time (ET), etc. (see Formula 3):

$$Y_{TM}^i = F\left(X_{EN}^i, X_{AN}^i, X_{ED}^i, X_{ET}^i\right) + \varepsilon_i \tag{3}$$

We used a quadratic function as the model regression function. The source data for each component included aggregated indicators of each student. For example: if we take the practical component, we deal with the sum of all the SCORM tasks marks, the time spent for all the tasks, the sum of the attempts to each task, the total number of the tasks. There have been calculated residuals for each of the regression factors, with help of a standard Excel Data Analysis module.

The obtained results (see Fig. 8) show experts' compliance in assessing the weights of the model edges and the constructed regression. Still some of the weights values need further consideration (for example, the time spent for a test). Besides, to make the model work more accurately there will be allocated additional entities and factors, that represent the course and the learning content more precisely.

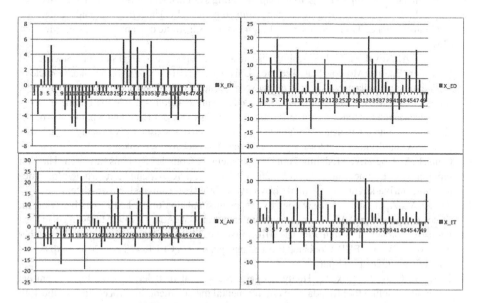

Fig. 8. The correlation and regression analysis results of an innovation marketing course.

6 Conclusion

Cognitive maps enable us to simulate weakly-formalized fields to improve prognosis quality and make up different scenarios of a situation. The article focuses on the possible ways to use fuzzy cognitive maps as a basis for modeling users' behavior patterns in the process of LMS e-learning. Further research will be focused on defining the proposed cognitive map parameters in terms of fuzzy functions that describe the map entities mutual influence, as well as the edges weights values.

References

1. Kosko, B.: Fuzzy cognitive maps. Int. J. Man Mach. Stud. **24**, 65–75 (1986)
2. Romero, C., Ventura, S., Garcia, E.: Data mining in course management systems: moodle case study and tutorial. Comput. Educ. **51**(1), 368–384 (2008)
3. Garcia, E., Romero, C., Ventura, S., de Castro, C.: A collaborative educational association rule mining tool. Internet High. Educ. **14**(2), 77–88 (2011)
4. E-Learning Market Trends & Forecast 2014–2016. https://www.docebo.com/landing/contactform/elearning-market-trends-and-forecast-2014-2016-docebo-report.pdf
5. Satokar, K.D., Gawali, S.: Web search result personalization using web mining. Int. J. Comput. Appl. **2**(5), 29–32 (2010). (0975–8887)
6. Ntarlas, O.D., Groumpos, P.P.: A survey on applications of fuzzy cognitive maps in business and management. Vestnik UGATU, 5 (2014). http://cyberleninka.ru/article/n/a-survey-on-applications-of-fuzzy-cognitive-maps-in-business-and-management
7. Axelrod, R.: Structure of Decision: The Cognitive Maps of Political Elite. Princeton University Press, Princeton (1976)
8. Kulinich, A.A.: Computer cognitive map modeling systems. Contr. Sci. **3**, 2–16 (2010)
9. Kulinich, A.A.: Semiotic cognitive maps (framed model). In: The XIIth National Conference on Control Science Transactions, pp. 4152–4164. RAS Institute of Control Sciences, Moscow (2014)
10. Uglev, V.A., Kovaleva, T.M.: Cognitive visualization to support individual studying. Science and Education: Bauman Moscow State Technical University, 3 (2014). http://cyberleninka.ru/article/n/kognitivnaya-vizualizatsiya-kak-instrument-soprovozhdeniya-individualnogo-obucheniya
11. Karelin, V.P.: Models and methods of knowledge representation and solution mining in fuzzy logic intellectual systems. Taganrog Institute of Management and Economy reporter, 1, vol. 19 (2014). http://cyberleninka.ru/article/n/modeli-i-metody-predstavleniya-znaniy-i-vyrabotki-resheniy-v-intellektualnyh-informatsionnyh-sistemah-c-nechyotkoy-logikoy
12. Ginis, L.A.: Cognitive Modeling Tool Development for Complex Systems Research. The Don reporter in Engineering, 3, vol. 26 (2013). http://cyberleninka.ru/article/n/razvitie-instrumentariya-kognitivnogo-modelirovaniya-dlya-issledovaniya-slozhnyh-sistem
13. Kireev, V.: Application of fuzzy cognitive maps in simulation of the LMS users behavior. In: Selected Papers of the XVIII International Conference on Data Analytics and Management in Data Intensive Domains (DAMDID/RCDL 2016), Ershovo, Moscow Region, Russia, 11–14 October 2016, vol. 1752 of CEUR Workshop Proceedings, pp. 65–69. CEUR-WS.org (2016). http://ceur-ws.org/Vol-1752/paper11.pdf
14. Gray, S., Gray, S., Cox, L., Henly-Shepard, S.: Mental modeler: a fuzzy-logic cognitive mapping modeling tool for adaptive environmental management. In: Proceedings of the 46th International Conference on Complex Systems, pp. 963–973 (2013)

Text Mining

Supporting Biological Pathway Curation Through Text Mining

Sophia Ananiadou[✉] and Paul Thompson

School of Computer Science, National Centre for Text Mining, University of Manchester, Manchester, UK
{sophia.ananiadou,paul.thompson}@manchester.ac.uk

Abstract. Text mining technology performs automated analysis of large document collections, in order to detect various aspects of information about their structure and meaning. This information can be used to develop systems that make it much easier for researchers to locate information of relevance to their needs in huge volumes of text, compared to standard search mechanisms. With a focus on the challenging task of constructing biological pathway models, which typically involves gathering, interpreting and combining complex information from a large number of publications, we show how text mining applications can provide various levels of support to ease the burden placed on pathway curators. Such support ranges from applications that provide help in searching and exploring the literature for evidence relevant to pathway reactions, to those which are able to make automated suggestions about how to construct and update pathway models.

Keywords: Text mining · Semantic search · Biological pathway curation · Event extraction · Named entity recognition

1 Introduction

Pathways are key to understanding biological systems. However, the construction of pathway models is dependent upon a complete and accurate representation of these systems, which requires that all relevant molecular species are captured, together with their physical interactions and chemical reactions.

Typically, complete information about the complex mechanisms involved in pathways is highly fragmented amongst many different information sources, making it necessary to assimilate knowledge from a plethora of heterogeneous sources, including not only the scientific literature, but also databases and ontologies. As an example, a manually constructed model of the *mToR* signaling network identified 964 species connected by 777 reactions, and reconstruction of this network required information to be gathered from a total of 522 publications [1].

The complex nature of reconstructing pathways involves not only *finding* appropriate information, but also evaluating, interpreting and distilling details from many different sources in order to construct a coherent and accurate model [2]. For example, it is not sufficient simply to locate statements in literature that provide evidence of relevant

© Springer International Publishing AG 2017

L. Kalinichenko et al. (Eds.): DAMDID/RCDL 2016, CCIS 706, pp. 59–73, 2017.
DOI: 10.1007/978-3-319-57135-5_5

reactions. Rather, it is also important to take into account the contextual details that accompany them (e.g., the degree of confidence or certainty expressed towards a finding) in order to determine their suitability for inclusion within the model. Furthermore, due to constant advances in scientific knowledge and the accompanying rapid growth in the literature, pathway models are not static objects. Instead, they must frequently be refined, verified and updated.

The various intricate stages of reconstructing pathway models typically require considerable amounts of manual effort by domain experts [3, 4]. Indeed, the slow and laborious nature of gathering and interpreting relevant information represents a major barrier to creating and maintaining pathway models, and the overwhelming volume of information that has to be reviewed can result in important details being overlooked.

In this article, we provide an overview of how text mining (TM) methods can support curators of pathways by allowing them to explore the vast body of scientific literature from a *semantic* perspective. This can significantly increase the ease and efficiency with which they can pinpoint, make sense of and use relevant information to construct and update pathways models. As such, TM has the potential to considerably increase the speed and reliability of knowledge discovery [5, 6].

2   Searching the Literature

In order to search for literature evidence that is relevant to support pathway curation, researchers will typically submit a query to a search engine, to try to retrieve a suitable set of publications. However, the querying mechanisms available in most search engines are poorly aligned with the information needs of the researcher.

The goal is usually to discover evidence about reactions (e.g., binding, phosphory-lation) that involve different types of concepts (e.g., genes/proteins, chemical compounds, subcellular components). In other words, the researcher is looking for textual evidence for highly specific pieces of *knowledge* that describe relationships amongst concepts.

In standard search engines, querying facilities are usually restricted to locating docu-ments that contain particular sets of words and/or phrases *somewhere* within them. However, isolated words and phrases do not in themselves convey knowledge. Docu-ments have highly complex structures, and knowledge is conveyed according to the *meaning* of phrases and how they are *related* to other phrases in the document.

Consider that a researcher wishes to discover proteins that bind to the protein *Mad1*. Using a standard search engine, a possible query would be *Mad1 AND bind*. Although this query is likely to retrieve *some* relevant documents, the fact that it is not possible to specify *how* these search terms should related to each other means that there is a very high probability of retrieving many documents where there is no specific rela-tion between the terms *Mad1* and *bind*. This implies a great deal of tedious "sifting" through irrelevant documents to isolate those of interest.

A further issue is that language usage is creative and unpredictable. Both concepts and the relationships in which they are involved can be expressed in text in a variety of different ways, using synonyms, abbreviations, acronyms, paraphrases, etc. For example, *Mad1* could appear in text as *MAD-1, mitotic arrest deficient-like protein 1* or *MAX dimerization protein 1*, while binding relationships could be expressed using words or phrases such as *recruit, forms a complex with*, etc. Trying to account for all such possible variants in a query would be virtually impossible, which leads to inevitable overlook of certain relevant documents.

Furthermore, many search terms can be ambiguous. For example, *ER* is a common abbreviation for *estrogen receptor*, but the same abbreviation may be used to refer to *endoplasmic reticulum* (a cellular subunit) and *emergency room*, amongst others. In the absence of sense-based filtering, searches for such terms will retrieve a large number of irrelevant documents.

3   Text Mining

Text mining methods can offer solutions to the issues above, by aiming to recognise various aspects of the semantic structure of documents [7]. TM accounts for the fact that (a) words and phrases have specific meanings, (b) particular meanings can be expressed in different ways, and (c) words and phrases can be related to each other in many different ways to convey complex information, such as reactions that are relevant to pathway curation. The results of TM analyses can be exploited in different applications that make it easier for researchers to pinpoint, filter and explore information that is of direct relevance to them. Such applications aim to minimise the tedious task of reading through irrelevant information, and reduce the likelihood of overlooking potentially vital information.

TM systems usually consist of a complex pipeline of different tools, which perform various levels of analysis that are required to gain an understanding of the information expressed in text. The functionalities of these tools range from low-level tasks, such as splitting a text into sentences and identifying the individual words within them, to more complex tasks, such as determining the parts-of-speech of individual words (e.g. nouns, verbs), grouping words into phrases and identifying structural (syntactic) relationships between these phrases, etc.

Typically, various tools are available that can carry out each individual task. This means that there are potentially many ways in which different tools could be combined to create TM processing pipelines. Since different tools may interact with each other in different ways, alternative pipelines with the same overall goal may perform with varying levels of accuracy. Accordingly, it can be advantageous to consider various combinations of different tools in order to construct an optimal TM system.

Until fairly recently, this could be a complex task, since different tools are implemented using various programming languages and have different input and output formats. The Argo TM platform [8] offers a solution to such issues, by providing a large (and continually growing) library of *interoperable* tools [9], which employ standardised input and output formats (based on the standards defined by the Unstructured Information Management Architecture (UIMA) [10]). This allows tools in the library to be flexibly combined into pipelines that carry out various different types of textual analysis (via a web-based, graphical user interface). Each pipeline can be evaluated and compared against others to find the best solution [11]. This makes it straightforward to build systems that perform various complex analyses of text, such as those described in the subsequent sections.

4   Concept Recognition

As outlined above, typical literature search goals revolve around discovering information about specific *concepts* rather than words. A researcher who searches for *estrogen receptor* will normally be interested in finding out information about a specific protein, regardless of the various ways in which it may be denoted in text (e.g. *oestrogen receptor, estrogen-receptor, ER* etc.).

The application of two well-established TM methods, i.e., named entity recognition (NER) and normalisation, can help to facilitate searching at the level of concepts rather than words. NER involves automatically identifying words and phrases in text that denote concepts of interest, and assigning a semantic label according to the concept category that they represent (e.g., *protein, chemical compound*, etc.). In the normalisation step, each phrase identified by NER is automatically *grounded* to a unique concept (usually by linking it to a domain specific database of known concepts). Accordingly, normalisation effectively identifies and links together all possible ways in which a concept could be expressed in text. Given the many potential variant expressions for a concept, combined with the fact that certain expressions may have multiple senses, the normalisation process may include disambiguation of acronyms according to their context [12] and comparison of phrases according to various types of surface-level and semantic similarities between them [13].

4.1   Kleio

The results of applying NER and normalisation to a collection of texts can form the basis of semantically-oriented search systems, such as Kleio[1] [14], which facilitates enhanced search over MEDLINE abstracts. In Kleio, semantic restrictions can be placed on query terms (such that, e.g., only documents in which *ER* refers to a protein are retrieved), and also to ensure that documents that use different ways of referring to the same concept are also automatically retrieved. This provides an important step towards reducing both overlook and overload of information.

Other functionality provided in Kleio further demonstrates the power of semantic TM analysis. Using NER and normalisation results for several different concept types, obtained by processing the whole of the MEDLINE, a *faceted* search mechanism makes it easier to explore and filter the results of an initial query according to their semantic content. Faceted search presents other concepts that frequently co-occur in the same documents as concepts of interest (e.g., proteins, symptoms, drugs, etc.). Frequent co-occurrences are likely to indicate interesting relationships.

---

[1]   http://www.nactem.ac.uk/Kleio/.

Figure 1 illustrates a search in Kleio for the protein *Mad1*. Opening the *Protein* facet reveals other proteins mentioned in the same documents as *Mad1*, the most common being *Mad2*. By selecting *Mad2* from the list, it is possible to "drill down" to documents in which the two proteins are mentioned together, and thus determine the nature of the relation that holds between them. The text snippets displayed, which mention a *Mad1–Mad2 complex*, provide evidence of their binding ability.

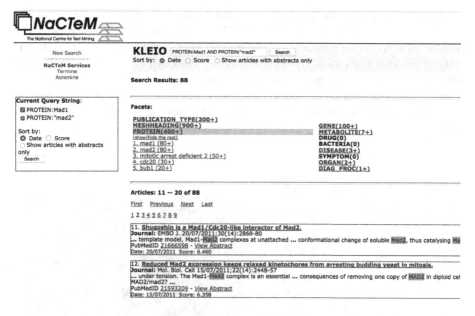

Fig. 1. Faceted search in Kleio

4.2   FACTA+

FACTA+[2,3] [15, 16] uses a type of TM analysis similar to Kleio, and has the aim of making it easy to find and visualise associations that occur between different concepts mentioned in MEDLINE abstracts. An additional useful feature of FACTA+ is the ability to specify that abstracts retrieved by a search should additionally contain mention of a *relationship* of a given type, e.g., *binding* or *positive regulation*, which can help to further filter the search results. A machine learning module determines the different ways in which each type of relation may be described. For example, searching for a binding relationship will automatically retrieve documents that include various different phrases that can denote binding, such as *recruitment, crosstalk, engagement, binding* and *inter-action*.

In FACTA+, it is possible to discover both directly associated concepts (i.e., concepts which are mentioned together in the same abstracts) and indirectly associated

---

[2] http://www.nactem.ac.uk/facta/.

[3] http://www.nactem.ac.uk/facta-visualizer/.

concepts. This latter functionality aims to account for the fact that associations may exist between concepts even if they are never mentioned together in the same document. The discovery of potential indirect associations works on the assumption that if two concepts *A* and *B* frequently occur together in documents, and if concept *B* frequently occurs together with a third concept, *C* (possibly in a distinct set of documents), then there is a possibility that concepts *A* and *C* also share some kind of association, via the "pivot" concept *B*.

Figures 2 and 3, respectively, illustrate direct and indirect associations in FACTA+ that are retrieved by a search for documents that mention *E-cadherin,* as well as a negative regulation relation. As can be seen in Fig. 2, the directly associated diseases are almost exclusively different forms of cancer. However, indirect associations (Fig. 3) reveal potential links with other diseases. For example, Parkinson's disease and Alzheimer's disease have a possible association with *E-cadherin* via the pivot gene concept *CASS4*. Thus, it could be hypothesised that E-cadherin may be a potential candidate drug target for nervous system disorders.

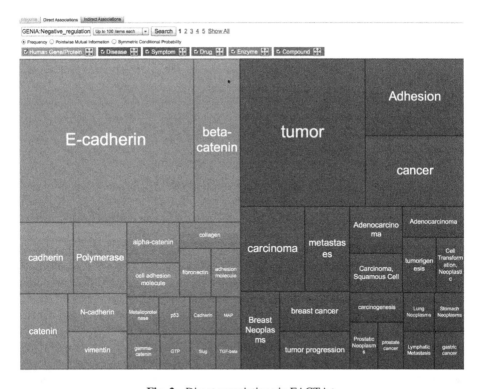

Fig. 2. Direct associations in FACTA+

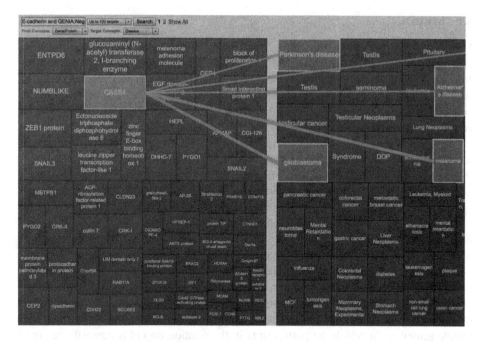

Fig. 3. Indirect associations in FACTA+

5   Relationship Recognition

Although systems such as Kleio and FACTA+ can make it far easier than a standard search engine to find potential relationships between concepts, they are still essentially "shallow" approaches, which only consider the semantics of individual words and phrases, without considering the exact nature of the relationship (if any) that holds between them. Whilst co-occurrence is used as an indicator of likely relationships, it is still necessary to carefully read the retrieved documents to determine whether a relationship of interest is actually described. For a complex task such as pathway reconstruction, in which many possible reactions have to be considered, this still implies a large amount of work on the part of the researcher.

Since constructing pathway models generally involves searching for very specific types of relationships, much work could be saved through the ability to identify more definite evidence of particular types of relationships that involve concepts of interest. However, this requires a more complex analysis of sentence structure, such that relationships between phrases can be identified.

5.1   MEDIE

MEDIE[4] [17] is a further search engine over MEDLINE abstracts, which combines NER [18] and normalisation with the results of a domain-specific syntactic parser [19]. MEDIE exploits the fact that in many cases, relationships between concepts are specified within the boundaries of a single sentence, and conveyed by means of a verb, where the concepts that constitute the "participants" in the relationship correspond to the syntactic arguments of the verb.

Syntactic analysis permits search queries in MEDIE to be *structured*. Instead of a search query consisting of a set of terms that must co-occur somewhere within documents, MEDIE allows specification of *how* the search terms should be related to each other. Queries in MEDIE are specified in terms of subject-verb-object triples – users can specify values for one or more of these slots, depending on the specificity of their query.

In Fig. 4, values have been specified for both the *subject* and *verb* slots, i.e., *p53* and *bind*, respectively, in order to discover what binds to p53. The query specifically finds sentences in which the word *bind* (or one of its inflections) occurs as a verb and *p53* occurs as its syntactic subject. In the retrieved results, the *object* phrase is highlighted in each relevant sentence, such that proteins that bind to p53 can be readily identified. Deep syntactic analysis allows MEDIE to find relationships described using a variety of sentence structures which may involve, e.g., the passive use of the verb. It can also handle sentences in which the participants in the relation do not necessarily occur in direct proximity to each other, e.g. *Furthermore, **p53** promotes cisplatin-induced apoptosis by directly **binding** and counteracting **Bcl-x(L)** antiapoptotic function.*

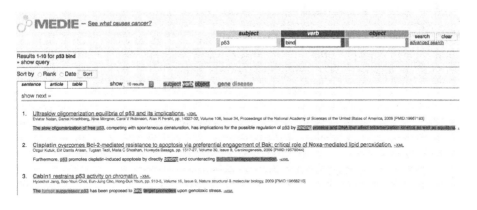

Fig. 4.   Structured searching in MEDIE

5.2   Beyond Simple Relations

Although syntactic analysis is a vital part of accurately determining relationships that hold between concepts, MEDIE can only find relationships that are centered around verbs. However, relationships described in other ways, particularly using

---

[4] http://www.nactem.ac.uk/medie/.

nominalisations, such *interaction* or *phosphorylation*, are highly prevalent in biomedical text [20]. Accordingly, failure to take such relationships into account risks overlooking potentially vital information.

Additionally, MEDIE is restricted to finding relations between *pairs* of concepts. However, such behaviour cannot account for the fact that, in constructing pathway models, it can be important to identify relations that involve multiple participants. For example, it can be vital to take into account the cellular context of a signaling event, such as cell type and localization; such information frequently occurs within the textual context of relationships.

Analysing the textual context of a relationship can also be important for other reasons. In particular, relationships can have different *interpretations*, e.g., they may not necessarily represent definite information. For example, in the sentence *We hypothesize that unphosphorylated cdr2 interacts with c-myc to prevent c-myc degradation*, the potential interaction between *cdr2* and *c-myc* is a hypothesis whose truth value is unknown. However, in building pathway models, it is important that only *reliable* knowledge is integrated into the model.

Whilst such contextual details could be found by careful reading of documents returned by a system such as MEDIE, the volume of documents that must typically be reviewed to construct pathway models means that researchers could benefit from systems that can automatically identify additional details about relationships. This can permit detailed information about the relationships to be presented to researchers, and/ or allow more complex filtering of relationships.

6 Event Recognition

Over recent years, a large amount of research has focused on the extraction of complex information structures from text, known as events (e.g., [21–27]). Importantly in the context of constructing pathway models, events can encode interactions that involve an arbitrary number of concepts, and are thus able to capture different types of contextual participants in interactions that are missing from binary relations.

Another important feature of events is their semantically-oriented representations of relationships, which abstract from the surface structure of the text. Firstly, events are assigned labels according to the general type of information that they convey (e.g., *negative regulation, phosphorylation, carboxylation*). Secondly, the participants in events are assigned specific *semantic role* labels (e.g., *modifier, reactant, product, cause, location*) according to their exact contribution towards the description of the relationship.

Generally, event extraction systems aim to capture all events of a given set of types that occur in a collection of documents, regardless of how they are expressed in text (e.g., using different nouns or verbs). This implies much more complex processing than the syntactic analyses carried out by MEDIE. Although syntactic analysis is still an important part of event extraction, the challenge lies in learning how to map from the surface level structure of the text to the more abstract semantic level event representation. For example, events of type *negative regulation* may be centred around verbs such as *inhibit* or *inactivate*, nouns such as *repression* or *loss*, or adjectives like *deficient* or *defective*. Depending on the exact word used, the mapping between syntactic structure and semantic participant roles may be different. For example, a possible way of denoting *positive regulation* events is using the verb *activate*, as in the following sentence: *Furthermore, a previous study has reported that **SAHA** activates p53*. Here, the *cause* of the positive regulation of *p53* is *SAHA*, which corresponds to the subject of the verb *activate*. However, other patterns will hold when the reaction is described by words belonging to other parts-of-speech. For example, if the positive regulation relationship is denoted by the noun *activation*, then it is likely that the *cause* will instead be preceded by the preposition *by*, i.e., **activation** *of p53 by* **SAHA**.

Despite the many complexities of recognising events automatically, they make it feasible to search for reactions based purely on a high-level semantic representation of the researcher's information needs, which is totally independent of the many ways in which the relevant knowledge may be expressed in documents. Specifically, event-based searching can allow the researcher to specify that they are looking for an event of a particular semantic category, and to place restrictions on the types of semantic participants being sought, without worrying about the many potential ways in which this information could be expressed in text.

6.1   EventMine

In order to extract complex events automatically, we have developed a pipeline-based event extraction system, EventMine [28], which employs a series of classifier modules to capture core event elements: detection of triggers (words or phrases that characterise the event), detection of edges (finding links between pairs of concepts), and complex event detection (combining edges into complex n-ary relations).

Since extracting event representations from text is heavily reliant on learning how to map from syntactic structure to semantic representations, EventMine uses a rich set of features, including those obtained from two different parsers [29, 30]. The system is also very flexible and can be adapted to extract different types of events without the need for task-specific tuning [31]. A further important feature of the EventMine is its incorporation of results from a pre-executed co-reference resolution method [32]. When event participants correspond to semantically empty expressions such as *it* and *that*, their exact interpretation is determined by looking in other sentences.

The results of applying EventMine to MEDLINE are used in an event-centric version of MEDIE (as illustrated in Fig. 5), in which search criteria are specified entirely in terms of structured semantic representations. Searches can be carried out over a number of different event types, and can place restrictions on participants that have different semantic roles. In Fig. 5, an event-based search for binding events involving *p53* shows that various different ways of expressing the binding relationship are recognised, in which the relation is specified using nouns as well as verbs.

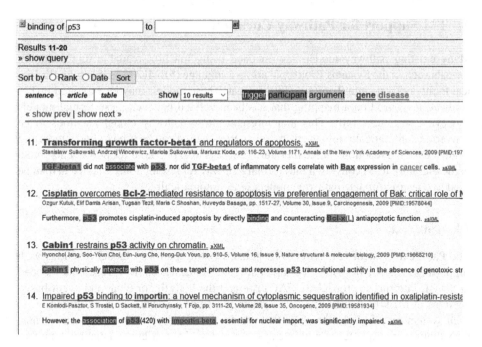

Fig. 5. Event-based querying in MEDIE

6.2 Event Meta-Knowledge

A recent enhancement to EventMine concerns the ability to detect and assign *interpretative* information to extracted events [33]. Values are determined for several different aspects or *dimensions* of interpretation, which we refer to collectively as *meta-knowledge* [34, 35]. By learning from a collection of texts in which event structures are manually annotated with meta-knowledge information [36], EventMine is able to automatically assign to each event its *polarity* (whether the event is negated), *knowledge type* (e.g., whether the event represents a well-known fact, a subject of investigation, an experimental observation or an analysis of experimental results), *certainty level, manner* (whether the reaction takes place with high or low intensity) and *source* (whether the event represents new or previously published information).

Such information can be useful in filtering events that describe potentially relevant interactions. For example, when constructing a new pathway model, it is likely to be important to consider *all* events that are considered sufficiently reliable, e.g., those that correspond both to well-known facts and experimental outcomes that are stated with a high degree of confidence, where the association is positive (rather than negated) and possibly excluding interactions that are characterised as being weak. In contrast, if the task is to *update* an existing pathway model to take into account the latest scientific knowledge, then the search may be further narrowed to consider only those events that represent new, confidently expressed experimental knowledge reported within articles published within a certain range of dates.

7   TM Support for Pathway Curation

Many existing pathway models are encoded using machine-readable representation formats such as the Systems Biology Markup Language (SBML) [37, 38] or the Biological Pathway Exchange (BioPAX) [39] format. These models may be exploited in TM applications, based on a mapping that has been defined between these formal models and event structures [40]. This mapping can allow pathway model reactions to be converted automatically into event-based queries, which can be used to find supporting evidence for the reactions in the literature. There is also potential for new events found in the literature to be converted into formal pathway representations, which can then be used to construct/update pathway models semi-automatically.

7.1   PathText[2]

The PathText[2] system [41] aims to associate existing pathway model reactions with supporting evidence from the literature. PathText[2] translates pathway reactions encoded in SBML into queries for Kleio, FACTA+ and the both the original and event-based versions of MEDIE. Given that each system identifies associations in a different way, each system may find relationships that are not extracted by the other systems. Thus, the submission of queries to multiple systems is aimed at retrieving a maximal number of documents that contain relevant evidence.

The results returned by each system are combined and presented to the user in a unified interface, ranked according to their relevance to pathway reactions (see Fig. 6). The ranking gives priority to the results of the event-based MEDIE, since experiments have shown this to be the most effective system in retrieving relevant documents. This further reinforces the importance of event extraction.

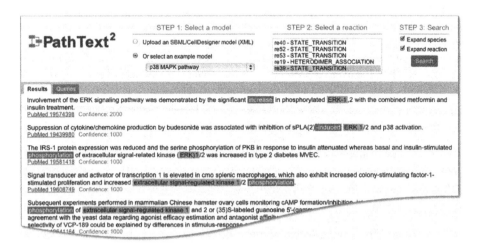

Fig. 6.   Literature evidence for pathway reactions in PathText[2]

7.2 Big Mechanism

The ability to link between formal pathway models and textual events is also being exploited in the *Big Mechanism* project[5], whose overall goal is to automate the process of intelligent, optimised drug discovery in cancer research. Pathway models will be used as the basis to generate new hypotheses for subsequent testing. TM techniques are being employed to construct, update and verify information in relevant models, to ensure that the information used for hypothesis generation is as accurate as possible. Events are extracted from the literature using EventMine, and are compared to event structures converted from reactions in existing pathway models.

The comparisons allow the existing models to be verified or updated in several ways. For example, events from the literature that match completely with events derived from the model act as corroborative evidence of the validity of these reactions. Other events found in the literature may help to extend the model, e.g., by identifying specific sites of a more general reaction, or by identifying a reaction that is not included in the model at all. By taking into account meta-knowledge, it is possible to identify potential contradictions for existing reactions in model. For example, a reaction included in the model may occur as a negated event in the literature, and thus it may require further investigation. Similarly, an existing reaction may be questioned if only tentative evidence for the reaction can be found in the literature.

8 Conclusion

Through their sophisticated semantic analysis of document collections, advanced TM methods can be used to develop search applications that can considerably increase the ease with which researchers can locate evidence to support tasks such as biological pathway curation, compared to traditional search methods.

In particular, the advances in the accuracy of automatic event extraction are paving the way for the development of systems that can immediately pinpoint information of direct relevance to the researcher and can largely eliminate the issues of information overload and overlook. Through the possibility of specifying information needs in terms of precise, abstract semantic structures, the burden of determining the many potential ways in which the information could be expressed in text can be increasingly shifted from the expert curator to the computer. According to the possibility of automated mapping between event structures extracted from text and formal pathway models, new opportunities are arising to further automate the processes of constructing, updating and validating pathway models. This will ultimately free experts from mundane, tedious tasks while aiding with more intellectually challenging ones.

Acknowledgements. The work described in this article has been supported by the BBSRC-funded *EMPATHY* project (Grant No. BB/M006891/1) and by the DARPA-funded *Big Mechanism* project Grant No. DARPA-BAA-14-14).

---

[5] http://www.nactem.ac.uk/big_mechanism/.

References

1. Caron, E., et al.: A comprehensive map of the mTOR signaling network. Mol. Syst. Biol. **6**, 453 (2010)
2. Oda, K., et al.: New challenges for text mining: mapping between text and manually curated pathways. BMC Bioinform. **9**(Suppl 3), S5 (2008)
3. Herrgard, M.J., et al.: A consensus yeast metabolic network reconstruction obtained from a community approach to systems biology. Nat. Biotechnol. **26**(10), 1155–1160 (2008)
4. Thiele, I., Palsson, B.Ø.: Reconstruction annotation jamborees: a community approach to systems biology. Mol. Syst. Biol. **6**, 361 (2010)
5. Ananiadou, S., McNaught, J. (eds.): Text Mining for Biology and Biomedicine. Artech House, Boston/London (2006)
6. Ananiadou, S., Kell, D.B., Tsujii, J.: Text mining and its potential applications in systems biology. Trends Biotechnol. **24**(12), 571–579 (2006)
7. Ananiadou, S.: Text mining bridging the gap between knowledge and text. In: Selected Papers of the XVIII International Conference on Data Analytics and Management in Data Intensive Domains (DAMDID/RCDL 2016), vol. 1752, pp. 140–141 (2016). http://ceur-ws.org/
8. Rak, R., et al.: Argo: an integrative, interactive, text mining-based workbench supporting curation. Database: J. Biol. Databases Curation **2012** (2012). bas010
9. Rak, R., et al.: Interoperability and customisation of annotation schemata in Argo. In: Proceedings of LREC, pp. 3837–3842 (2014)
10. Ferrucci, D., et al.: Towards an interoperability standard for text and multi-modal analytics. IBM Research Report RC24122 (2006)
11. Batista-Navarro, R., Rak, R., Ananiadou, S.: Optimising chemical named entity recognition with pre-processing analytics, knowledge-rich features and heuristics. J. Cheminf. **7**(Suppl. 1), S6 (2015)
12. Okazaki, N., Ananiadou, S., Tsujii, J.: Building a high-quality sense inventory for improved abbreviation disambiguation. Bioinformatics **26**(9), 1246–1253 (2010)
13. Alnazzawi, N., Thompson, P., Ananiadou, S.: Mapping phenotypic information in heterogeneous textual sources to a domain-specific terminological resource. PLoS ONE **11**(9), e0162287 (2016)
14. Nobata, C., et al.: Kleio: a knowledge-enriched information retrieval system for biology. In: Proceedings of the 31st Annual International ACM SIGIR, pp. 787–788 (2008)
15. Tsuruoka, Y., Tsujii, J., Ananiadou, S.: FACTA: a text search engine for finding associated biomedical concepts. Bioinformatics **24**(21), 2559–2560 (2008)
16. Tsuruoka, Y., et al.: Discovering and visualizing indirect associations between biomedical concepts. Bioinformatics **27**(13), i111–i119 (2011)
17. Miyao, Y., et al.: Semantic retrieval for the accurate identification of relational concepts in massive textbases. In: Proceedings of ACL, pp. 1017–1024 (2005)
18. Tsuruoka, Y., Tsujii, J.: Bidirectional inference with the easiest-first strategy for tagging sequence data. In: Proceedings of HLT/EMNLP, pp. 467–474 (2005)
19. Hara, T., Miyao, Y., Tsujii, J.: Adapting a probabilistic disambiguation model of an HPSG parser to a new domain. In: Dale, R., Wong, K.-F., Su, J., Kwong, O.Y. (eds.) IJCNLP 2005. LNCS (LNAI), vol. 3651, pp. 199–210. Springer, Heidelberg (2005). doi: 10.1007/11562214_18
20. Cohen, K.B., Palmer, M., Hunter, L.: Nominalization and alternations in biomedical language. PLoS ONE **3**(9), e3158 (2008)
21. Kim, J.-D., et al.: Extracting bio-molecular event from literature—The BioNLP'09 shared task. Computational Intelligence **27**(4), 513–540 (2011)

22. Kim, J.-D., Pyysalo, S., Nedellec, C., Ananiadou, S., Tsujii, J. (eds.): Selected Articles from the BioNLP Shared Task 2011. BMC Bioinformatics, vol. 13, Suppl. 11 (2012)

23. Nédellec, C., Kim, J.-D., Pyysalo, S., Ananiadou, S., Zweigenbaum, P. (eds.): BioNLP Shared Task 2013: Part 1. BMC Bioinformatics, vol. 16, Suppl. 10 (2015)

24. Nédellec, C., Kim, J.-D., Pyysalo, S., Ananiadou, S., Zweigenbaum, P. (eds.): BioNLP Shared Task 2013: Part 2. BMC Bioinformatics, vol. 16, Suppl. 16 (2015)

25. Thompson, P., Iqbal, S., McNaught, J., Ananiadou, S.: Construction of an annotated corpus to support biomedical information extraction. BMC Bioinform. **10**, 349 (2009)

26. Pyysalo, S., et al.: BioInfer: a corpus for information extraction in the biomedical domain. BMC Bioinform. **8**, 50 (2007)

27. Ananiadou, S., et al.: Event-based text mining for biology and functional genomics. Brief. Funct. Genomics **14**(3), 213–230 (2015)

28. Miwa, M., et al.: Event extraction with complex event classification using rich features. J Bioinform. Comput. Biol. **8**(1), 131–146 (2010)

29. Sagae, K., Tsujii, J.: Dependency parsing and domain adaptation with LR models and parser ensembles. In: Proceedings of the CoNLL 2007 Shared Task, pp. 1044–1050 (2007)

30. Miyao, Y., et al.: Evaluating contributions of natural language parsers to protein-protein interaction extraction. Bioinformatics **25**(3), 394–400 (2009)

31. Miwa, M., Ananiadou, S.: Adaptable, high recall, event extraction system with minimal configuration. BMC Bioinform. **16**(Suppl. 10), S7 (2015)

32. Miwa, M., Thompson, P., Ananiadou, S.: Boosting automatic event extraction from the literature using domain adaptation and coreference resolution. Bioinformatics **28**(13), 1759–1765 (2012)

33. Miwa, M., et al.: Extracting semantically enriched events from biomedical literature. BMC Bioinform. **13**, 108 (2012)

34. Nawaz, R., et al.: Meta-knowledge annotation of bio-events. Proc. LREC **2010**, 2498–2507 (2010)

35. Nawaz, R., Thompson, P., Ananiadou, S.: Evaluating a meta-knowledge annotation scheme for bio-events. In: Proceedings of the Workshop on Negation and Speculation in Natural Language Processing, pp. 69–77 (2010)

36. Thompson, P., et al.: Enriching a biomedical event corpus with meta-knowledge annotation. BMC Bioinform. **12**, 393 (2011)

37. Hucka, M., et al.: The systems biology markup language (SBML): a medium for representation and exchange of biochemical network models. Bioinformatics **19**(4), 524–531 (2003)

38. Hucka, M., et al.: Evolving a lingua franca and associated software infrastructure for computational systems biology: the Systems Biology Markup Language (SBML) project. Syst. Biol. **1**(1), 41–53 (2004)

39. Demir, E., et al.: The BioPAX community standard for pathway data sharing. Nat. Biotechnol. **28**(9), 935–942 (2010)

40. Ohta, T., Pyysalo, S., Tsujii, J.: From pathways to biomolecular events: opportunities and challenges. In: Proceedings of BioNLP 2011 Workshop, pp. 105–113 (2011)

41. Miwa, M., et al.: A method for integrating and ranking the evidence for biochemical pathways by mining reactions from text. Bioinformatics **29**(13), i44–i52 (2013)

Text Processing Framework for Emergency Event Detection in the Arctic Zone

Dmitry Devyatkin and Artem Shelmanov[(✉)]

Federal Research Center "Computer Science and Control" of Russian Academy of Sciences,
Moscow, Russia
{devyatkin,shelmanov}@isa.ru

Abstract. We present the text processing framework for detection and analysis of events related to emergencies in a specified region. We consider the Arctic zone as a particular example. The peculiarity of the task consists in data sparseness and scarceness of tools/language resources for processing such specific texts. The system performs focused crawling of texts related to emergencies in the Arctic region, information extraction including named entity recognition, geotagging, vessel name recognition, and detection of emergency related messages, as well as indexing of texts with their metadata for faceted search. The framework aims at processing both English and Russian text messages and documents. We report the results of the experimental evaluation of the framework components on Twitter data.

Keywords: Focused crawling · Event detection · Text stream monitoring · Named entity recognition · Emergency message detection · Vessel name recognition · Faceted search

1 Introduction

During emergency situations, a lot of data are generated in social networks, blogs, mass media, and other sources. The substantial part of these data is unstructured natural language texts like short messages, news articles, official reports. As shown by many researchers, these texts contain valuable information that could be useful both for the affected people and for the emergency response and rescue teams. Leveraging these unstructured data requires the development of frameworks and methods that are adapted to heterogeneous nature of these data and peculiarities of tasks that arise in emergency event monitoring. Currently, there is a lack of techniques and tools suitable for building a framework for such task, which makes this problem difficult. The problem becomes more challenging if additionally we need to build a system that is focused not just on a particular topic but also on a particular geographical region.

This paper continues the study reported in [1]. It presents a framework and methods for intelligent processing of textual data for monitoring and detection of emergency events in the Arctic zone. The important peculiarity of this task lies in the restricted amount of available data sources, as well as in lack of ready-to-use software, methods and linguistic resources for processing such specific data. The developed framework

© Springer International Publishing AG 2017
L. Kalinichenko et al. (Eds.): DAMDID/RCDL 2016, CCIS 706, pp. 74–88, 2017.
DOI: 10.1007/978-3-319-57135-5_6

primarily is intended for processing English and Russian natural language texts. The following tasks are the focus of the framework:

- Focused crawling of data related to emergency events in the Arctic zone. Although the everyday information stream is extremely large, the volume of topically relevant information is relatively small. Therefore, it is very inefficient and almost impossible to store and index all the available data. The preliminary filtering is required that would sift irrelevant documents and substantially reduce the amount of stored and indexed data. The potential information sources of the developed framework are not restricted to only mass media and official reports of emergency response and rescue teams, but they also can include messages from social networks like Twitter, Facebook, Vkontakte, and others.
- Information extraction from natural language texts for recognizing and tagging important objects. The extracted information can be used for text classification and for result filtering during the information search.
- Faceted search in the indexed database of the crawled information stream.
- Detection of important events in the crawled information stream, including emergency events, evaluation of the descriptiveness and importance of texts and messages.

In the current paper, we propose several methods and approaches for solving most of the aforementioned tasks.

In theory, it seems that the problem of focused crawling for emergency events in a geographical region could be solved by methods of topic crawling, in practice, we need to crawl the texts related to the variety of topics that also change in time. We propose focused crawling method for gathering relevant information about Arctic zone from heterogeneous information sources that uses multiple different types of crawlers. These crawlers implement various crawling and filtering strategies for improving recall and precision of the gathered data.

Creating tools for information extraction from texts requires many natural language resources. However, the amount of such resources that are related to the Arctic zone is very limited (especially for the Russian language), which complicates application of common machine learning techniques. Therefore, we adapt open information extraction paradigm, and use pre-trained word embeddings, as well as the publicly available corpora and databases to create the information extraction pipeline in the framework. In this paper, we discuss three major functions implemented in the information extraction pipeline: location tagging, vessel name recognition, detection of emergency related messages.

The faceted search provides a mechanism for filtering information search results by multiple feature groups like topic, time, location, relations to some objects, etc. For faceted search in the developed framework, we propose extensive usage of methods of linguistic analysis including syntax and semantic parsing (semantic role labeling and relation extraction), as well as annotating and tagging of important objects and metainformation in texts like geographical coordinates and named entities (organizations, persons, locations, vessel names). Metainformation and annotations are indexed with text and could be used in the multifaceted search.

We experimentally evaluate the proposed methods on several test datasets created using Twitter crawling. Although this social network accumulates only short messages and is not designed for providing data for the considered tasks, many researchers, as shown in Sect. 2, demonstrated that tweets could be a useful source of information about emergencies and used it as a common benchmark. Therefore, we use messages crawled from Twitter for evaluation of the system components. However, we note that the developed framework is designed to handle all sorts of textual information, not just short messages.

This paper extends [1] with a review of state-of-the-art methods relevant for the problem in consideration (Sect. 2), with a presentation of methods for detection of emergency related messages (Subsect. 3.2), for vessel name recognition (Subsect. 3.2), and for focused crawling using extracted vessel names (Subsect. 3.1), as well as with some additional experimental results (Subsect. 4.2). The paper is structured as follows. Section 2 presents the review of the state-of-the-art methods and systems for monitoring of emergency events in social media messages. Section 3 describes the proposed methods of focused crawling of messages related to emergency events in the Arctic zone, natural language processing and information extraction pipeline, and the method for the faceted search. In Sect. 4, we present the results of experimental evaluation of the proposed methods. Section 5 concludes and discusses the future work.

2   Related Work

In recent works, the task of emergency event detection via text processing is primarily related to analyzing of message streams of social networks. Large-scale emergency events give rise to a massive publication activity in social networks [2]. These message streams contain information about situation in the affected area, infrastructure damage, casualties, the requests and proposals for help. This is crucial information that can enhance the situation awareness of affected people and emergency response teams involved in rescue operations [3]. There are several systems related to the problem of emergency event detection by monitoring messages in social media. Most of the known systems deal primarily with short Twitter messages in English. In [1], we briefly reviewed several of them including ESA [3], Twitris [4], SensePlace 2 [5], EMERSE [6], AIDR [7], Tweedr [8], TEDAS [9]. The aforementioned systems are mostly oriented on large-scale emergencies (e.g., earthquakes) and are not spatially restricted. Many of them specialize on narrow problems like message classification, whereas our research is oriented on the development of a full-stack framework that solves many tasks: from focused crawling and information extraction, to faceted search leveraged with spatial and temporal metadata. In this aspect, our work is close to [10]. In this work, the complex platform for emergency event detection is presented. The architecture of the platform provides several components that perform data capturing and filtering, event detection, emergency monitoring, which is mostly related to assessment of damage and severity of an event, and alert dissemination (aggregating data for a final user). However, in this work, authors propose a platform that is oriented on customization to certain types of emergencies without restrictions on geographical region, whereas our framework is aimed at monitoring as much as possible types of emergencies in a specified region, which complicates focused crawling and filtering of data. This platform is aimed at deeper analysis of emergency events, whereas we consider information search as the primary function of our framework.

The review shows that systems related to emergency event detection help to improve situational awareness in crisis situations by means of focused crawling, classification of messages into informative/non-informative ones, topical clustering of messages for better search, as well as extraction of spatial and temporal information from texts. In the rest of the section, we review the state-of-the-art methods for some of such tasks that are relevant for our work.

There are two main issues that arise in the task of disaster message crawling: the shifts of target topics over the time and limitations of social networks APIs. Among many previous works related to the first problem, we should note [11]. In this study, researchers present a method for sustainable topical crawling in Twitter social network. The method periodically extracts keywords from downloaded messages and uses them for shifting of target topics over the time. The first problem also could be tackled by method implemented in the iCrawl framework – a tool for creating focused crawling systems [12]. This framework uses the global full-text search engines for extraction of the original seeds for focused crawling. In [13], the method for distributed crawling of retrospective text collections is proposed that tackles the second aforementioned problem. The method is able to collect data from different groups of users and overcome some limitations of the Twitter search API. Thus, we combine these approaches in our study.

Among previous studies related to disaster message classification, we should note [6]. In this work, authors present the approach to extraction of valuable messages about emergency events in social media. For this task, they used naïve Bayesian and maximum-entropy classifiers. Besides some low-level features (words and n-grams) authors also synthesized several complex high-level features like a target audience of a message (personal or impersonal) and a message format (formal or informal). In the conducted experiments, they showed that using the high-level features greatly improves the performance of the classifiers. Different approach was used in [14]. Instead of synthesizing complex features, researchers used neural network for disaster tweet classification task. They showed that their models based on neural network that do not require manual feature engineering perform better than state-of-the-art methods. The proposed method allows the accurate processing of the out-of-event data and reaches good results in the early hours of a disaster, when no labeled data is available. In [15], an approach to detection of useful messages in social media streams during disaster events is presented. Authors used convolutional neural networks that showed significant improvement in performance over models that use the traditional "bag of words" and n-grams as features on several datasets of messages related to flooding events.

Focused crawling, as well as the faceted search require some metainformation about texts including named entities. The majority of the successful approaches to named entity recognition (NER) use supervised machine learning on annotated corpora. However, in many languages we observe the lack of the annotated corpora sufficient for training classifiers based on supervised techniques. There is still no such openly available corpus for Russian language. In such cases, the methods based on semi-supervised open information extraction approaches can provide a solution for this task.

In [16], the language agnostic method for named entity recognition was presented. The method does not need human-annotated corpus for learning, instead it automatically generates it using Wikipedia and Freebase [17]. In the proposed method, the spans of Wikipedia articles tagged with a hyperlink to another Wikipedia article are marked as entity annotations. Only terms that are present in the entity articles of Freebase are considered as training examples. The Freebase is also used to determine the category of annotations. Such an approach generates corpus with some false-negatives. Authors tackle this problem by presenting additional oversampling and surface word matching methods. The learning process is performed in a language agnostic manner using only word embeddings that encode semantic and syntactic features of words in different languages, and are built by a neural network. The method was used for creating NER parser for 40 languages including Russian. In [18], a new Twitter entity disambiguation dataset was proposed, and empirical analysis of named entity recognition and disambiguation was conducted. The researchers investigated robustness of a number of state-of-the-art systems on noisy messages, the main sources of their errors, and other related problems. In [19], researchers proposed a new corpus of Weibo social network messages annotated for evaluating NER. They also evaluated three types of neural embeddings for representing Chinese text and proposed a joint objective for training embeddings that makes use of both labeled and unlabeled texts. The proposed methods yield substantial improvement over the baseline.

Most of the state-of-the-art machine learning approaches for classification and NER require large labeled datasets that are specific to a particular disaster. An elaborated feature engineering is also needed to achieve best results. Previous works empirically showed that neural network based models do not require any feature engineering and perform better than other methods on many tasks. Therefore, we adopt methods based on state-of-the-art neural network architectures for detection of emergency related and for NER in our work.

In the research area devoted to faceted search, [20, 21] are worth noting. These works show the limitations of the classic approaches to faceted search in the way the data, facets, and queries are modeled. It is shown that these limitations could be tackled by usage of semantic web technologies. Another way to outperform the classic approaches was proposed in [22]. The paper presents a search engine wherein deep NLP framework is successfully used to improve the search accuracy. The faceted search in our framework is also based on advanced search engine that leverages the deep linguistics information including syntax relations, semantic roles, and other types of semantic annotations extracted from natural language texts.

3   Methods for Focused Crawling, Text Processing, and Faceted Search Implemented in the Framework

3.1   Focused Crawling

Monitoring emergency situations in the Arctic zone first of all requires document crawling from heterogeneous sources: social networks like Twitter, Facebook, VKontakte and news aggregators like ArcticInfo, BarentsObserver, BBC, etc. Text documents and messages can include geographical coordinates or geotags that could be used for position location of their authors and filtering of the crawled stream. Some sources are entirely devoted to the Arctic zone; thus, they can be crawled without additional filtering by location. Concerning that, for crawling, we use several types of crawlers.

The first type is geotag crawlers. They gather messages with geographical coordinates or geotags. During the message stream processing, these crawlers filter out messages, which geotag latitude is less than 60°.

The second type is topic crawlers. These crawlers are intended to download topical relevant messages without geotags and coordinates. Each topic is specified by two sets: relevant keywords and key phrases and stop-words that should not be present in the downloaded messages. These sets are used for topic search via Twitter API. The main difficulty lies in limitations of Twitter search API. Besides that, it is not very expressive and is very restrictive to query length (maximum 500 URL-encoded characters). All these limitations have serious negative impact both on the recall and on the precision of the crawling process. The problem with insufficient recall can be tackled by creating multiple topic crawlers configured with different subsets of keywords and key phrases. To improve the precision, we suggest extending the crawling seeds with entities automatically extracted from downloaded messages. Named entities usually are more informative, therefore, using them results in crawling data that are more relevant to the target topic. Another approach to improving precision of topic crawlers is using ship names as seeds. When a ship name is detected in a message, approximate location of the ship is determined by automated identification system (AIS). The information about the ship location is provided by open-access Internet resources that publish AIS data. The acquired spatial information is used for filtering messages by geographical region, whereas the ship name can be added as another seed term for crawling in the next iterations.

Another type of crawlers is indented for gathering information from sources that yield only topically relevant information: topically relevant section of news aggregators, the archives of the government services, and specific blogs in social networks. These crawlers do not need additional filtering.

The algorithm of data crawling is the following. In the first step, messages are collected by geotag and topic crawlers. In the second step, we apply natural language processor and information extraction pipeline to the collected texts. Then, we filter all messages that do not contain toponyms and geotags, or are not related to emergency events. Fuzzy duplicates of messages effectively are filtered using inverted text indexes [23]. URLs from the remaining messages are fed into the common crawler that processes only relevant sources. For topic crawler, we build a topic model [24] of the crawled messages every several days. It helps to track topic shifts in the message stream. We summarize topic content with a keyword cloud and a set of the most significant messages from the cluster. Then each topic is marked as relevant or irrelevant by several assessors. The most significant terms from the relevant topics are sent to "permissive" keyword collections of topic crawlers, and terms from irrelevant topics are sent to "restrictive" ones. Thus, the crawling process becomes responsible to trend shifts.

3.2 Natural Language Processing and Information Extraction

The system performs deep natural language processing of Russian and English texts. Besides basic processing, the pipeline also performs syntax parsing, semantic role labelling, and information extraction.

The basic analysis for Russian texts is performed by AOT.ru[1]. This framework is used for tokenization, sentence boundary detection, POS tagging, and lemmatization, including morphological disambiguation. We use MaltParser[2] trained on SynTagRus [25] for dependency parsing of Russian texts and our semantic parser for semantic role labelling [26]. The same types of linguistic analysis of English texts are performed via Freeling [27]. Note that the syntax and semantic annotations are used for information search (see Sect. 3.3).

For the basic named entity recognition, we use Polyglot NER framework [16]. It produces annotations for locations, organizations, and person names. However, we found that the basic NER processor is not suitable for extracting toponyms related to a particular region (e.g., Arctic zone); it yields low recall in this task. Therefore, we complemented Polyglot with a gazetteer.

The gazetteer was created on the basis of Geonames[3] database. It contains more than 11 million geographical locations around the world with their names (in many languages including Russian and English), geographical coordinates, and other metadata. From Geonames, we extracted location names that are situated on the north of the 60th latitude. The gazetteer uses these data to mark spatial information in texts. It also implements rather simple rules to filter out common false positives that take into account parts of speech and capitalization of words.

Another important task related to emergency event detection in the Arctic zone consists in monitoring events that involve sea and river vessels, since the big part of the Arctic zone is covered by water. Therefore, the information extraction pipeline in our framework includes a recognizer of vessel names in texts. This information is used in the method of focused crawling (see Sect. 3.1) and in the faceted search.

The vessel name recognizer is based on another gazetteer and machine learning techniques. The gazetteer dictionary was created on the basis of the open database of sea and river vessels[4]. Besides names, the database contains a vast amount of information and metadata about ships including flags of vessels.

---

[1] http://aot.ru/.

[2] http://maltparser.org/.

[3] http://www.geonames.org/.

[4] http://www.marinevesseltraffic.com/.

The preliminary experiments showed that using only gazetteer yields low precision for the task of vessel name recognition. The big part of false positives was due to the fact that many vessel names are general use words and location names. However, in most cases, the context can help to solve this ambiguity. To diminish the fraction of false positives and improve the precision we use a classifier based on machine learning. In particular, we use the small recurrent neural network with long-short term memory cells (LSTM) [28]. Figure 1 displays the architecture of the network. The architecture is based on C-LSTM model proposed in [29]. The purpose of the network consists in refining the results of the gazetteer with binary classification. The input of the network is a sequence of word embeddings. We use embeddings that were preliminarily built applying GloVe to a massive corpus of Twitter messages [30]. The decision to use pre-trained embeddings instead of adding an embedding layer into the network directly is caused by a small amount of available training data. The input layer of the neural network is convolutional. It performs one-dimensional convolutions of trigrams. It is followed by max-pooling and a simple LSTM recurrent layer. The final processing is done by two-layer dense network with a linear hidden layer and a binary output with sigmoid activation. Note that, the network processes words from a whole sentence with a vessel annotation, and the vessel annotation itself is replaced by a special token.

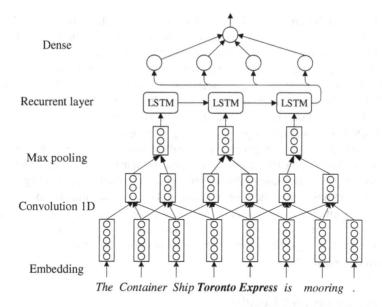

Fig. 1. The architecture of the neural network for vessel name recognition and detection of emergency related messages

For training, we collected a small balanced corpus of tweets. True positives were collected using specific vessel names that clearly represent a vessel in a text message, as well as keywords related to sea topic. These messages were processed by the gazetteer and the obtained results were refined manually by annotators. Partially, false positives were collected automatically by crawling tweets using general-purpose lexis as seeds.

These messages also were annotated by gazetteer but in this case all annotations were automatically marked as negative. We got 1500 samples for training and another 500 samples for hyper-parameter tuning.

The text processing and information extraction pipeline also provides an ability to detect messages related to emergency events. We tag messages related to emergency situations using lexicon CrisisLex[5], proposed in [31]. Although such tagging is useful for text search and focused crawling, such an approach also produces many false positives due to ambiguity of natural language. To solve this problem, we adopt a similar approach used for vessel name recognition. In this task, neural network is applied directly to the whole message and classifies it into emergency or not emergency. As in previous task, we used neural network with a convolutional layer, max pooling, a LSTM recurrent layer, and two dense layers. However, in this task preliminary experiments showed that non-linear hidden dense layer produces better results. To collect training set, we used CrisisLex as a seed for topic crawlers. The crawled messages were manually annotated. For the training of the neural network, we used 2500 samples, another 500 samples were used for hyper-parameter tuning.

3.3 Faceted Search

The faceted search became a backbone for professional search applications [32]. In this type of search, users can iteratively specify queries using metadata and keywords extracted from search results of previous iterations. Additionally, search results could be filtered using different sets of meta fields that can be static or dynamic.

In the developed framework, the faceted search is powered by the Exactus technology [33]. Its main advantage lies in ability to efficiently index rich linguistic information including syntax relations, semantic roles, and other types of semantic annotations extracted from natural language text (e.g. named entities). This enables phrase search (results have to contain given syntactically connected phrases) semantic search (results are ranked taking into account semantic similarity of the query and indexed documents). We take advantage of this technology by introducing indexing by geographical tags, timestamps, emergency-related tags, and other tags. This provides the ability to filter results efficiently by semantic information like location, time, organizations, persons, vessels, emergency tag, and topics. It also provides the ability to retrieve information with certain tags filtered by other metadata producing the results that can be sifted with consequent queries.

4 Experiments and Results

We conducted a series of experiments to assess the performance of the created components for focused crawling, information extraction, and faceted search. We mostly used Twitter social network as the source of experimental data. This approach was chosen because many systems in previous works are tested in the similar way – on the basis of

---

[5] http://crisislex.org/.

Twitter data, since this social network often becomes an important informational channel in emergency situations for large segments of population.

4.1 Focused Crawling Evaluation

To assess the accuracy of the proposed focused crawling framework, we evaluated the quality of filtering on the downloaded datasets. We crawled all topic-relevant messages and tweets with appropriate geographical coordinates or geotags. These messages are marked as "Impure data". We labeled several subsets of the received posts devoted to accidents in Alaska and Bering Sea. Each post from the subsets was labelled by three assessors to reach sufficient coherence of the test data. In the annotated data, approximately 2 k messages were marked as relevant to emergency events. We processed all these messages by the proposed NER and the detector of emergency messages. And finally, we filtered messages, which do not contain mentions of any location, person, organization, or an emergency event marked by the baseline CrisisLex gazetteer. In these experiments, we did not use information about vessels and results of emergency event detector based on neural network. We have not applied a cross-validation approach here because the labelling was not used for the crawler training, just for testing. The standard measures for supervised learning: precision, recall, and F_1-score, were used for each subset. Macro-averaging was used to evaluate the result assessments. Table 1 refers to the results of the crawling without and with filtering as "Impure data" and "Filtered data" correspondingly.

Table 1. Focused crawling evaluation results

	P	R	F_1
Impure data	0.26	1.00	0.41
Filtered data	0.57	0.94	0.70

Results show that applying the proposed filtering technique yields substantial growth of the precision without the significant decrease of the recall. This means that during the crawling process we do not lose much topically relevant data but substantially decrease the stored noise. We decided to choose a fairly soft filtering because, although a stricter procedure would improve the precision, it would also imply a more significant recall drop, which contradicts the purpose of the monitoring framework.

4.2 Information Extraction Evaluation

We evaluated performance of several components of the information extraction pipeline: the parser for location tagging, the vessel name recognizer, and the detector of emergency messages.

In the experiments with the parser for location tagging, we wanted to evaluate a general purpose named entity recognizer in a specific domain (in Arctic zone in particular) and assess the importance of usage of the specialized language resources. We compared the

Polyglot NER with the developed gazetteer, and with the combined processor, in which results of both tools are merged.

We prepared the test set that contains 300 geographical locations in Twitter messages related to Arctic zone. The achieved results are presented in Table 2.

Table 2. Location tagging evaluation results

Parser	P	R	F_1
Polyglot	0.78	0.57	0.66
Gazetteer	0.78	0.74	0.76
Polyglot + Gazetteer	0.76	0.82	0.79

Results show that the proposed gazetteer significantly outperforms Polyglot on location extraction in terms of recall. The knowledge source of Polyglot is Wikipedia and Freebase that do not have the full coverage of locations. We conclude that it is reasonable to use the gazetteer and Polyglot together for the maximum performance.

We tested the vessel name recognizer using a holdout set consisted of 500 tweets. The results of the evaluation are presented in Table 3. Since the test examples were created using the gazetteer and neural network was used only for refinement, we considered that the recall of the gazetteer on the test set is ideal and its precision is just a fraction of true positives in the test set. The goal of this evaluation was to compare the performance of the baseline direct approach based on the gazetteer with the performance of the developed neural network, assess the quality of filtering performed by neural network, and assess the drop of the recall of vessel name recognition in this case.

Table 3. The results of the vessel name recognition evaluation

Parser	P	R	F_1
Gazetteer	0.41	1.00	0.58
Neural network	0.90	0.92	0.91

The results show that using the neural network leads to relatively small drop of the recall but also to big improvement in the precision. Thus, the proposed approach provides a decent tradeoff between precision and recall yielding substantially higher F_1-score than the baseline.

The detector of emergency related messages was evaluated in the similar way. We used the holdout consisted of 500 tweets as a test set. It was composed by a baseline – the search of keywords from CrisisLex. The recall of the baseline on the test set is ideal and the precision is a fraction of false positives marked by annotators. The results of the evaluation are presented in Table 4.

Table 4. The evaluation results of emergency related message detection

Parser	P	R	F_1
Baseline	0.42	1.00	0.59
Neural network	0.84	0.89	0.86

The neural network yields much better F_1-score than the baseline (27% improvement). The neural network is superior by 42% in terms of the precision. However, it also shows a 11% recall drop. While precision improvement is good, for practical use the tradeoff should be tuned on behalf of better recall, since it is more important. This could be done by tuning the threshold of the classifier.

4.3   Faceted Search Evaluation

We assessed the performance gain of the information search achieved by using the proposed emergency faceted search method in comparison to the baseline algorithm. In the experiment, we used more than 100 k crawled messages from Twitter. They were indexed by Exactus search engine. As the baseline, we deployed our full-text search algorithm without filtering by tag locations. For the evaluation, we applied the widely used nDCG score and peer-review approach. The results of the evaluation are presented in Table 5.

Table 5.   Faceted search evaluation results

	3-DCG	5-DCG	10-DCG
Baseline	0.61	0.55	0.53
Faceted	0.76	0.76	0.70

It was revealed that use of location and extracted emergency tags for faceted search significantly improves the quality of ranking when searching for posts about emergencies.

5   Conclusion

In this paper, we presented the ongoing work on the text processing framework for emergency event detection in a specified geographical region (Arctic zone, in particular) and proposed the methods for solving the major tasks that arise in this field. We addressed the problems of focused crawling, information extraction, text classification, and faceted search.

For focused crawling, we suggested to use multiple types of crawlers that rely on different approaches including topic modeling, as well as usage of spatial and other types of information extracted from downloaded texts. To build the information extraction pipeline we used Polyglot, gazetteers, and recurrent neural networks. For the training of the neural networks, we created two corpora: corpus of tweets annotated with sea and river vessel names and corpus of tweets that are manually classified into emergency-related and neutral. The natural language processing pipeline of our framework also includes syntax and semantic parsing. Syntax, semantics, and other information extracted from texts are indexed by the Exactus search engine, which provides faceted search in the crawled text stream. Thus, our framework provides a coherent stack of methods and tools for solving the task of emergency event detection.

We conducted a series of experimental studies, in which we compared the performance of the proposed methods with the baseline approaches. The obtained results are promising and demonstrate that the proposed framework could be useful on practice for monitoring events in text streams restricted by a geographical region.

In the future work, we are going to extend our information extraction pipeline with new features. In particular, we are going to incorporate into the pipeline methods for evaluation of informativity of social network messages. This feature could be useful for calculating relevance in faceted search, as well as for constructing more informative seed terms for focused crawler. We also are looking forward to take more advantage of temporal information stored in text messages.

We are going to accumulate more retrospective data from social networks and other sources to increase the recall of the crawling process. Among many other types of information sources, collections of reports from rescue services are the most prospect supplement for the crawling. Another way to improve crawling is detection of users and groups in social networks that constantly post topically relevant messages. This could be done semi-automatically by building topical models for users and groups. We are also going to create visualization tools for geotagged messages that can present events on a map.

Acknowledgments. The project is supported by the Russian Foundation for Basic Research, project number: 15-29-06045 "ofi_m".

References

1. Deviatkin, D., Shelmanov, A.: Towards text processing system for emergency event detection in the Arctic zone. In: Selected Papers of the XVIII International Conference on Data Analytics and Management in Data Intensive Domains, CEUR Workshop Proceedings, pp. 148–154 (2016)
2. Sixto, J., Pena, O., Klein, B., López-de Ipina, D.: Enable tweet-geolocation and don't drive ERTs crazy! Improving situational awareness using Twitter. In: Proceedings of SMERST, pp. 27–31 (2013)
3. Yin, J., Karimi, S., Robinson, B., Cameron, M.: ESA: emergency situation awareness via microbloggers. In: Proceedings of the 21st ACM International Conference on Information and Knowledge Management, pp. 2701–2703. ACM (2012)
4. Purohit, H., Sheth, A.P.: Twitris v3: from citizen sensing to analysis, coordination and action. In: Proceedings of ICWSM, pp. 746–747 (2013)
5. MacEachren, A.M., Jaiswal, A., Robinson, A.C., Pezanowski, S., Savelyev, A., Mitra, P., Zhang, X., Blanford, J.: Senseplace2: Geotwitter analytics support for situational awareness. In: Proceedings of Visual Analytics Science and Technology (VAST) on IEEE Conference, pp. 181–190 (2011)
6. Verma, S., Vieweg, S., Corvey, W.J., Palen, L., Martin, J.H., Palmer, M., Schram, A., Anderson, K.M.: Natural language processing to the rescue? Extracting "situational awareness" tweets during mass emergency. In: Proceedings of ICWSM, pp. 385–392 (2011)
7. Imran, M., Castillo, C., Lucas, J., Meier, P., Vieweg, S.: AIDR: artificial intelligence for disaster response. In: Proceedings of the Companion Publication of the 23rd International Conference on World Wide Web Companion, pp. 159–162 (2014)
8. Ashktorab, Z., Brown, C., Nandi, M., Culotta, A.: Tweedr: mining Twitter to inform disaster response. In: Proceedings of ISCRAM, pp. 354–358 (2014)

9. Li, R., Lei, K.H., Khadiwala, R., Chang, K.C.C.: Tedas: a Twitter-based event detection and analysis system. In: 2012 IEEE 28th International Conference on Data engineering (ICDE), pp. 1273–1276. IEEE (2012)

10. Avvenuti, M., Del Vigna, F., Cresci, S., Marchetti, A., Tesconi, M.: Pulling information from social media in the aftermath of unpredictable disasters. In: 2015 2nd International Conference on Information and Communication Technologies for Disaster Management (ICT-DM), pp. 258–264. IEEE (2015)

11. Li, R., Wang, S., Chang, K.C.C.: Towards social data platform: automatic topic-focused monitor for Twitter stream. Proc. VLDB Endowment **6**(14), 1966–1977 (2013)

12. Gossen, G., Demidova, E., Risse, T.: The iCrawl wizard – Supporting interactive focused crawl specification. In: Hanbury, A., Kazai, G., Rauber, A., Fuhr, N. (eds.) ECIR 2015. LNCS, vol. 9022, pp. 797–800. Springer, Cham (2015). doi:10.1007/978-3-319-16354-3_88

13. Boanjak, M., Oliveira, E., Martins, J., Mendes Rodrigues, E., Sarmento, L.: Twitterecho: a distributed focused crawler to support open research with twitter data. In: Proceedings of the 21st International Conference Companion on World Wide Web, pp. 1233–1240. ACM (2012)

14. Nguyen, D.T., Mannai, K.A.A., Joty, S., Sajjad, H., Imran, M., Mitra, P.: Rapid classification of crisis-related data on social networks using convolutional neural networks (2016). arXiv preprint arXiv:1608.03902

15. Caragea, C., Silvescu, A., Tapia, A.H.: Identifying informative messages in disaster events using convolutional neural networks. In: International Conference on Information Systems for Crisis Response and Management (2016)

16. Al-Rfou, R., Kulkarni, V., Perozzi, B., Skiena, S.: Polyglot-NER: massive multilingual named entity recognition. In: Proceedings of the 2015 SIAM International Conference on Data Mining, SIAM (2015)

17. Bollacker, K., Evans, C., Paritosh, P., Sturge, T., Taylor, J.: Freebase: a collaboratively created graph database for structuring human knowledge. In: Proceedings of the 2008 ACM SIGMOD International Conference on Management of Data, pp. 1247–1250. ACM (2008)

18. Derczynski, L., Maynard, D., Rizzo, G., van Erp, M., Gorrell, G., Troncy, R., Petrak, J., Bontcheva, K.: Analysis of named entity recognition and linking for tweets. Inf. Process. Manage. **51**(2), 32–49 (2015)

19. Peng, N., Dredze, M.: Named entity recognition for Chinese social media with jointly trained embeddings. In: EMNLP, pp. 548–554. Association of Computational Linguistics (2015)

20. Arenas, M., Cuenca Grau, B., Evgeny, E., Marciuska, S., Zheleznyakov, D.: Towards semantic faceted search. In: Proceedings of the 23rd International Conference on World Wide Web, pp. 219–220. ACM (2014)

21. Bast, H., Buchhold, B.: An index for efficient semantic full-text search. In: Proceedings of the 22nd ACM International Conference on Information & Knowledge Management, pp. 369–378. ACM (2013)

22. Armentano, M.G., Godoy, D., Campo, M., Amandi, A.: NLP-based faceted search: experience in the development of a science and technology search engine. Expert Syst. Appl. **41**(6), 2886–2896 (2014)

23. Zubarev, D., Sochenkov, I.: Using sentence similarity measure for plagiarism source retrieval. In: CLEF (Working Notes), pp. 1027–1034 (2014)

24. Hofmann, T.: Probabilistic latent semantic indexing. In: Proceedings of the 22nd Annual International ACM SIGIR Conference on Research and Development in Information Retrieval, pp. 50–57. ACM (1999)

25. Nivre, J., Boguslavsky, I.M., Iomdin, L.L.: Parsing the SynTagRus treebank of Russian. In: Proceedings of the 22nd International Conference on Computational Linguistics (Coling 2008), pp. 641–648 (2008)

26. Shelmanov, A.O., Smirnov, I.V.: Methods for semantic role labeling of Russian texts. In: Computational Linguistics and Intellectual Technologies, Papers from the Annual International Conference "Dialogue", no. 13, pp. 607–620 (2014)

27. Padró, L., Stanilovsky, E.: Freeling 3.0: towards wider multilinguality. In: Proceedings of the Language Resources and Evaluation Conference (LREC 2012). ELRA (2012)

28. Hochreiter, S., Schmidhuber, J.: Long short-term memory. Neural Comput. **9**(8), 1735–1780 (1997)

29. Zhou, C., Sun, C., Liu, Z., Lau, F.: A C-LSTM neural network for text classification (2015). arXiv preprint arXiv:1511.08630

30. Pennington, J., Socher, R., Manning, C.D.: GloVe: global vectors for word representation. EMNLP **14**, 1532–1543 (2014)

31. Olteanu, A., Castillo, C., Diaz, F., Vieweg, S.: CrisisLex: a lexicon for collecting and filtering microblogged communications in crises. In: Proceedings of ICWSM (2014)

32. Fafalios, P., Tzitzikas, Y.: Exploratory professional search through semantic post-analysis of search results. In: Paltoglou, G., Loizides, F., Hansen, P. (eds.) Professional Search in the Modern World. LNCS, vol. 8830, pp. 166–192. Springer, Cham (2014). doi: 10.1007/978-3-319-12511-4_9

33. Osipov, G., Smirnov, I., Tikhomirov, I., Sochenkov, I., Shelmanov, A.: Exactus expert – search and analytical engine for research and development support. In: Hadjiski, M., Kasabov, N., Filev, D., Jotsov, V. (eds.) Novel Applications of Intelligent Systems, vol. 586, pp. 269–285. Springer, Switzerland (2016)

Fact Extraction from Natural Language Texts with Conceptual Modeling

Mikhail Bogatyrev[✉]

Tula State University, Tula, Russia
okkambo@mail.ru

Abstract. The paper presents the application of Formal Concept Analysis paradigm to the fact extraction problem on natural language texts. Proposed technique combines the usage of two conceptual models: conceptual graphs and concept lattice. Conceptual graphs serve as semantic models of text sentences and the data source for concept lattice – the basic conceptual model in the Formal Concept Analysis. With the use of concept lattice it is possible to model relationships between words from different sentences from different texts. These relationships have been collected in formal concepts of concept lattice and provide interpreting formal concepts as possible facts. Facts can be extracted by using navigation in the lattice and interpretation its concepts and hierarchical links between them. Experimental investigation of the proposed technique is performed on the annotated textual corpus consisted of descriptions of biotopes of bacteria.

Keywords: Conceptual graphs · Formal Concept Analysis · Concept lattice

1 Introduction

The problem of fact extraction from text is the part of more general problem of knowledge extraction from text [1]. Methods for solving this problem are strongly depended on whether the text is structured or not. We will use the term "text" for natural language text and the term "textual data" when text is structured by means of database or corpus. Facts and events form a kind of knowledge which represents semantics of a certain portion of text. In this area of research the term "event" is applied in the literature more often than "fact" [2] and sometimes these terms have similar meaning. However we distinguish facts and events in the corresponding problems of knowledge extraction.

Both facts and events extracted from texts can be represented by words and relationships on the sets of words. An example of fact is phrase "SAP has purchased SYBASE" and this phrase also denotes an event of purchasing. The model of this event may be in the form of pattern *<agent>*-purchase-*<patient>* where concrete words may be substituted as semantic roles of agent and patient. In the survey [2] facts are defined as "statistical relations", so the evidence of facts is detected statistically and discovered relations "are not necessarily semantically valid, as semantics (meanings) are not explicitly considered, but are assumed to be implicit in the data" [2]. Now this definition may be replenished so that relations in the fact model can be found semantically valid and the evidence of facts is detected not statistically but also semantically. Certain

© Springer International Publishing AG 2017
L. Kalinichenko et al. (Eds.): DAMDID/RCDL 2016, CCIS 706, pp. 89–102, 2017.
DOI: 10.1007/978-3-319-57135-5_7

technologies, including one presented in this paper, devoted to extract facts using semantics explicitly presented in corresponding semantic models of text. Many of these models are the same as in the fact extraction problems as in the event extraction problems: for example lexico-syntactic and lexico-semantic patterns are applied there. These models are also applied for solving Named Entity Recognition (NER) problem. Solutions of this problem often come as the base for solutions of fact extraction problem [2].

We consider fact as realized or occurred event. So the modeling of events and facts may be implemented in a same way. We apply conceptual modeling [5] in the fact extraction problem. This method is based on the usage of two conceptual models, conceptual graphs and concept lattice, to discover facts as formal concepts and their relationships in concept lattice.

Conceptual modeling is one of the ways of modeling semantics in the Natural Language Processing (NLP) [6, 7]. Every conceptual model has its own semantics which represents the meanings of concepts and relationships on them.

Formal Concept Analysis (FCA) [17] is the paradigm of conceptual modeling which studies how objects can be hierarchically grouped together according to their common attributes. In the FCA, its conceptual model is the lattice of formal concepts (concept lattice) which is built on the abstract sets treated as objects and their attributes. Concept lattices have been applied as an instrument for information retrieval and knowledge extraction in many applications. The number of FCA applications now is growing up including applications in social science, civil engineering, planning, biology, psychology and linguistics [22, 23]. Several successful implementations of FCA methods on fact extraction on textual data [12, 13] and Web data are known [19]. Although the high level of abstraction makes FCA suitable for use with data of any nature, its application to specific data often requires special investigation. It is fully relevant for using FCA on textual data.

Another paradigm of conceptual modeling is Conceptual Graphs [25]. Conceptual graph is bipartite directed graph having two types of vertices: concepts and conceptual relations. Conceptual terms of entities and relationships are represented in conceptual graphs as its concepts and conceptual relations.

Conceptual graphs have been applied for modeling many real life objects including texts. Acquiring conceptual graphs from natural language texts is non-trivial problem but it is quite solvable [5].

There is great number of various methods of solving fact and event extraction problems which can be distinguished according to data-driven and knowledge-driven approaches [2]. Data-driven approach is based on the idea that knowledge (facts or events) presented explicitly in data whereas knowledge-driven approach requires external resources or expert knowledge for solving the problem.

Fact extraction technology proposed in this paper is hybrid. Using conceptual graphs as semantic model of text we follow the data-driven approach. Expert knowledge-driven methods are applied in the output of the technology when facts have to be detected and presented in the output interface. The principles of creating this technology are described in [6] and its implementation in biomedical data research is described in [9]. In this paper we present some generalizations of these principles and new experimental results of investigation of biotopes of bacteria.

2   Fact Extraction Technology

The work of fact extraction technology is illustrated on the Fig. 1.

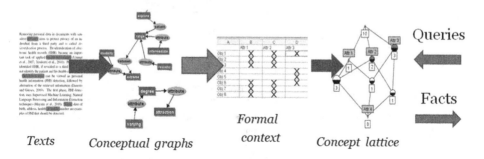

Texts   *Conceptual graphs*   *Formal context*   *Concept lattice*

Fig. 1. Elements of the fact extraction technology.

The elements of this technology have the following content.

1. Input data in the form of plain text is transformed to the set of conceptual graphs. The maximal number of conceptual graphs is equal to the number of processed sentences of texts.
2. According to FCA paradigm, so called formal context is building on the set of conceptual graphs. It is a matrix denoting a relation on two sets of objects and their attributes. These sets must be determined on the set of conceptual graphs. This stage is a crucial step in the technology. The number of formal contexts and their content depends on many factors and is domain-specific.
3. Formal context contains formal concepts which are combinations of objects and attributes that meet certain conditions known as Galois connection and constitute a lattice named as concept lattice [17]. Concept lattice is interpreted as storage of facts. Facts can be extracted by processing input textual queries and then navigating in the lattice and interpreting its concepts and hierarchical links between them.

2.1   Acquiring and Implementing Conceptual Graphs

The method of acquiring conceptual graphs from natural language texts is considered in [5]. Some peculiarities of conceptual graphs created with this method are illustrated in [6, 7].

The method has standard phases of lexical, morphological and semantic analysis extended with the solution of the problem of semantic role labeling [8]. This problem is non-trivial since semantic roles do not belong to the sentence processed and must be discovered from existing roles by means of morphological analysis.

Semantic analysis on the stage of acquiring conceptual graphs is domain-specific. For example, working with biological domain and not considering its specificity we will not acquire correct conceptual graph for the following sentence:

   "HI2424 is characterized as a representative of the B. cenocepacia PHDC clonal lineage".

Wrong conceptual graph is a graph which has isolated concepts do not linked with any other concepts as it is shown on Fig. 2.

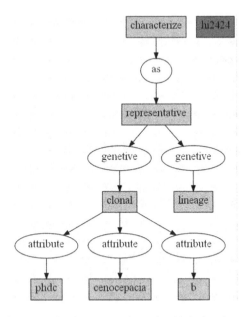

Fig. 2. Example of conceptual graph with isolated concept.

For the sentence above and for similar sentences being characteristic for biological domain we use supervised learning and external resources in the form of textual corpus. Then, after learning, the algorithm of acquiring conceptual graphs knows that *B. cenocepacia* is a shortcut name of the *Burkholderia cenocepacia* bacterium, *HI2424* is the code of this bacterium and *PHDC* is the name of the clone of bacteria.

Extracting facts is performed on the same stage of creating conceptual graphs. Some isolated concepts appearing on applying the algorithm before its learning may indicate facts. Figure 2 illustrates this showing conceptual graph for the sentence discussing above before the algorithm learning. Here the presence HI2424 code in the sentence is the fact that marks this sentence as having information about *Burkholderia cenocepacia bacterium* which will be used later to filter non-informative sentences.

The next stage of the fact extraction technology is creating formal contexts and concept lattice as the main conceptual model serving as the source of facts. Conceptual graphs and FCA models are closely related when they are applied as conceptual models in text processing. One of the first mentioning of this relation is in [30]. Now it is used in connection with the problem of aggregation of conceptual graphs.

2.2 Conceptual Graphs and Formal Concept Analysis

There are two basic notions FCA deals with: *formal context* and *concept lattice* [17]. Formal context is a triple $\mathbf{K} = (G, M, I)$, where G is a set of objects, M – set of their attributes, $I \subseteq G \times M$ – binary relation which represents facts of belonging attributes to objects. The sets G and M are partially ordered by relations $\sqsubseteq$ and $\in$, correspondingly: $G = (G, \sqsubseteq)$, $M = (M, \in)$. Formal context is represented by [0, 1] matrix $\mathbf{K} = \{k_{i,j}\}$ in which units mark correspondence between objects $g_i \in G$ and attributes $m_j \in M$. The concepts in the formal context have been determined by the following way. If for subsets of objects $A \subseteq G$ and attributes $B \subseteq M$ there exist mappings (which may be functions also) $A': A \to B$ and $B': B \to A$ with properties of $A': = \{\exists m \in M \mid < g, m > \in I \; \forall g \in A\}$ and $B': = \{\exists g \in G \mid < g, m > \in I \; \forall m \in B\}$ then the pair (A, B) that $A' = B$, $B' = A$ is named as *formal concept*. The sets A and B are closed by composition of mappings: $A'' = A$, $B'' = B$; A and B is called the *extent* and the *intent* of a formal context $\mathbf{K} = (G, M, I) \in$, respectively.

By other words, a formal concept is a pair (A, B) of subsets of objects and attributes which are connected so that every object in A has every attribute in B, for every object in G that is not in A, there is an attribute in B that the object does not have and for every attribute in M that is not in B, there is an object in A that does not have that attribute.

The partial orders established by relations $\sqsubseteq$ and $\in$ on the set G and M induce a partial order $\leq$ on the set of formal concepts. If for formal concepts (A_1, B_1) and (A_2, B_2), $A_1 \sqsubseteq A_2$ and $B_2 \in B_1$ then $(A_1, B_1) \leq (A_2, B_2)$ and formal concept (A_1, B_1) is less general than (A_2, B_2). This order is represented by *concept lattice*. A lattice consists of a partially ordered set in which every two elements have a unique *supremum* (also called a least upper bound or *join*) and a unique *infimum* (also called a greatest lower bound or *meet*).

According to the central theorem of FCA [17] a collection of all formal concepts in the context $\mathbf{K} = (G, M, I)$ with subconcept-superconcept ordering $\leq$ constitutes the *concept lattice* of $\mathbf{K}$. Its concepts are subsets of objects and attributes connected each other by mappings A', B' and ordered by a subconcept-superconcept relation.

To illustrate these abstract definitions consider an example. Figure 3 shows simple formal context and concept lattice composed on the sets $G = \{DNA, Virus, Prokaryotes, Eukaryotes, Bacterium\}$ and $M = \{Membrane, Nucleus, Replication, Recombination\}$. The set G is ordered according to sizes of its elements: DNA is smallest and bacterium is biggest ones. The set M has relative order: one part (*Membrane, Nucleus*) characterizes microbiological structure of objects from G, but another part (*Replication, Recombination*) characterizes the way of breeding, and these parts are incomparable.

	Membrane	Nucleus	Replication	Recombination
DNA				X
Virus	X			X
Prokaryotes	X		X	
Eukaryotes	X	X	X	
Bacterium	X		X	

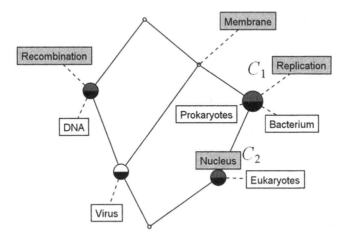

Fig. 3. Example of formal context and concept lattice.

The lattice for formal context on Fig. 3 is drawn compact and is interpreted in the following way. There are empty concepts on the top and on the bottom of the lattice diagram. Every formal concept lying on the path from top to bottom contains attributes (shown dark) which are gathered from the concepts lying before. Vice versa, every formal concept lying on the path from bottom to top contains objects (shown bright) which are gathered from the concepts lying before. That is why the concept $C_1 = (\{Prokaryotes, Eukaryotes, Bacterium\}, \{Membrane, Replication\})$ contains the object *Eukaryotes* and the attribute *Membrane*. The concept C_1 is more general than the concept $C_2 = (\{Eukaryotes\}, \{Membrane, Replication, Nucleus\})$.

Also on the Fig. 3 there is the fact of existing two different branches of concepts characterizing two families: $\{viruses, DNA\}$ and $\{prokaryotes, eukaryotes, bacteria\}$. The link between them is the attribute "*Membrane*". It is known [11] that viruses can have a lipid shell formed from the membrane of the host cell. Therefore, the membrane is positioned in the formal context on the Fig. 3 as an attribute of the virus.

This example demonstrates specific ways of extracting knowledge from conceptual lattice:

- analyzing formal concepts in concept lattice;
- analyzing conceptual structures in concept lattice – its paths and sub lattices in the general case.

These ways are applied in our previous [9] and current research of bacteria biotopes.

FCA on Textual Data. The main problem in applying FCA to textual data is the problem of building formal context. If textual data is represented as the natural language texts then this problem becomes acute. There are several approaches to the construction of formal contexts on the textual data, presented as separate documents, as data corpora. One, mostly applied variant is the context in which the objects are text documents and the attributes are the terms from these documents. Another variant is building formal context directly on the texts and the formal context may represent various features of textual data:

- semantic relations (synonymy, hyponymy, hypernymy) in a set of words for semantic matching [20],
- verb-object dependencies from texts [14],
- words and their lexico-syntactic contexts [21, 24].

These lexical elements must be distinguished in texts as objects and attributes. There are following approaches to solve this problem:

- creating corpus tagging by adding special descriptions in texts which mark objects and attributes [10],
- using semantic models of texts [14].

We apply the second approach and use conceptual graphs for representing semantics of individual sentences of a text.

Aggregation of Conceptual Graphs and Pattern Structures. In the theory of conceptual graphs aggregation means replacing conceptual graphs by more general graphs [25]. These general graphs may be created as new graphs or may be graphs or sub graphs from initial set of graphs. Aggregation of conceptual graphs has semantic meaning and general graphs make up the context (not formal context) of initial set of graphs.

One way of aggregation is conceptual graphs clustering. Graphs which are the nearest ones to the centers of clusters have been treated as general graphs.

We have studied several approaches for clustering conceptual graphs using various similarity measures [6] and applied clustering for creating formal concepts on conceptual graphs.

Another way of conceptual graphs aggregation is based on supporting types of concepts of conceptual graphs. Types of concepts have been implied in the model of conceptual graph [17]. To support types of concepts, external resources are needed. They may be thesaurus or textual corpus with tagging or ontology.

According to generalization of FCA [16] conceptual graphs and their external resource may be considered as pattern structures.

2.3 Creating Formal Contexts with Conceptual Graphs

The crucial step in the described process of CGs – FCA modeling is creating formal contexts on the set of conceptual graphs.

At the first glance, this problem seems simple: those concepts of conceptual graphs which are connected by "attribute" relation have been put into formal context as its objects and attributes. Actually the solution is much more complex.

To provide the presence information about those and other facts in the formal contexts the following rules are implemented as mostly important when creating formal contexts.

1. Not only individual concepts and relations, but also patterns of connections between concepts in conceptual graphs represented as sub graphs have been analyzed and processed. These patterns are the predicate forms *<object>* - *<predicate>* - *<subject>* which in conceptual graphs look as the template *<concept>* - *(patient)* - *<*verb*>* - *(agent)* - *<concept>*. Not only *agent* and *patient* semantic roles but also other roles are allowed in the templates.
2. The hierarchy of conceptual relations in conceptual graphs is fixed and taken into account when creating formal context. Using this hierarchy of conceptual relations it is possible to select for formal contexts more or less details from conceptual graphs.

These empirical rules are related to the principle of pattern structures which was introduced in FCA in the work [16]. A pattern structure is the set of objects with their descriptions (patterns), not attributes. Patterns also have similarity operation. The instrument of pattern structures is for creating concept lattices on the data being more complicated than sets of objects and attributes.

Conceptual graph is a pattern for the object it represents. A sub graph of conceptual graph is projection of a pattern. Namely projections are often used for creating formal contexts. Similarity operation on conceptual graphs is a measure of similarity which is applied in clustering.

3   Fact Extraction from Biomedical Data

Bioinformatics is one of the fields where Data Mining and Text Mining applications are growing up rapidly. New term of "Biomedical Natural Language Processing" (BioNLP) has been introduced there [4]. This term is stipulated by huge amount of scientific publications in Bioinformatics and organizing them into corpora with access to full texts of articles via such systems as PubMed [26]. Information resources of PubMed have been united in several subsystems presenting databases, corpora and ontologies.

So called "research community around PubMed" [18] forms data intensive domain in this area. It not only uses data from PubMed but also creates new data resources and data mining tools including specialized languages for effective biomedical data processing [15].

In our experiments we use PubMed vocabulary thesaurus MeSH (Medical Subject Headings) as external resource for supporting types of concepts in conceptual graphs.

3.1 Data Structures

Our experiments have been carried out using text corpus of bacteria biotopes which is used in the innovation named as BioNLP Shared Task [10]. Biotope is an area of uniform environmental conditions providing a living place for plants, animals or any living organism. Biotope texts form tagged corpus. The tagging includes full names of bacteria, its abbreviated names and unified key codes in the database. We can add additional tags and we do it.

A BioNLP data is always domain-specific. All the texts in the corpus are about bacteria themselves, their areal and pathogenicity. Not every text contains these three topics but if some of them are in the text then they are presented as separate text fragments. This simplifies text processing.

The fact extraction technology is realized as experimental modeling framework [7] having DBMS for storing and managing data used in experiments. We use relational database on the SAP-Sybase platform. Database stores texts, conceptual graphs, formal contexts and concept lattices. Special indexing is applied on textual data.

3.2 BioNLP Tasks

According to the BioNLP Shared Task initiative [10] there are two main tasks solving on biomedical corpora: the task of Named Entity Recognition (NER) and the task of Relations Extraction (RE).

The task of Named Entity Recognition on the corpus of bacteria descriptions is formulated as seeking bacteria names presented directly in the texts or as co-references (anaphora).

Relations Extraction means seeking links between bacteria and their habitat and probably diseases it causes. The task of Named Entity Recognition has direct solution with conceptual graphs. The only problem which is here is anaphora resolution.

Anaphora resolution is the problem of resolving references to earlier or later items in the text. These items are usually noun phrases representing objects called referents but can also be verb phrases, whole sentences or paragraphs. Anaphora resolution is the standard problem in NLP.

Biotopes texts we work with contain several types of anaphora:

- hypernym defining expressions ("bacterium" - "organism", "cell" - "bacterium"),
- higher level taxa often preceded by a demonstrative determinant ("this bacteria", "this organism"),
- sortal anaphoras ("genus", "species", "strain").

For anaphora detection and resolution we used a pattern-based approach. It is based on fixing anaphora items in texts and establishing relations between these items and bacteria names. Additional details may be found in [6, 9].

3.3 Fact Extraction with Concept Lattices

Conceptual graphs represent relations between words. Therefore they can be applied for relations extraction but only in one sentence. For extracting relations between bacteria in several texts we applied concept lattices.

We had selected 130 mostly known bacteria and have processed corresponding corpus texts about them. All the texts were preliminary filtered for excluding stop words and other non-informative lexical elements.

Three formal contexts of "Entity", "Areal" and "Pathogenicity" were built on the texts. They have the names of bacteria as objects and corresponding concepts from conceptual graphs as attributes. Table 1 shows numerical characteristics of created contexts.

Table 1. Numerical characteristics of created contexts.

Context name	Number of objects	Number of attributes	Number of formal concepts
Entity	130	26	426
Areal	130	18	127
Pathogenicity	130	28	692

Among attributes there are bacteria properties (gram-negative, rod-shaped, etc.) for "Entity" context, mentions of water, soil and other environment parameters for "Areal" context and names and characteristics of diseases for "Pathogenicity" context.

As it is followed from the table there is relatively small number of formal concepts in the contexts. This is due to the sparse form of all contexts generated by conceptual graphs.

Visualization in Fact Extraction. Visualization plays significant role in FCA [28] and in fact extraction since not only formal concepts but also relations between concepts in a concept lattice may be treated as facts, and visualization helps to find them fast. But it allows getting results only for the relatively small lattices. For extracting facts we use visualization together with other ways including database technologies. A possibility was created to visualize sub lattices of a concept lattice to form special views constructed on the lattice corresponding to certain property (intent in the lattice) or entity (extent in the lattice) on the set of bacteria. We applied open source tool [29] which was modified and built in our system.

Consider the example demonstrating the work of the system. One of the problems solving in investigations of bacteria biotopes is the problem of bacteria classification: it is needed to classify bacteria according to their properties characterizing them as the entities, characterizing their areal and pathogenicity. Various bacteria may have similar properties or may not. It is interesting to find clusters of bacteria containing ones having similar properties. This clustering task may be solved with a concept lattice.

Figure 4 shows a fragment of the formal context with the attributes related to some properties of bacteria: Gram staining, the property of being aerobic, etc.

	gram-negative	Gram-positive	acid-fast	rod-shaped	spiral	aerobic	anaerobic	non-spore	bacteria
Mycobacterium bovis		X	X	X		X			X
Mycoplasma agalactiae		X							
Bifidobacterium longum		X						X	
Corynebacteria		X							
Mycobacteria		X	X	X					X
Streptomycetes		X							X
Mycobacterium tuberculosis		X				X			
Clostridium		X					X		X
Clostridium sticklandii		X					X		
Deinococcus radiodurans		X							
Deinococcus deserti	X								
Cupriavidus metallidurans	X							X	
Thermoanaerobacter tengcongensis	X			X			X		
Thermus thermophilus	X					X			
Sinorhizobium meliloti	X					X			
Bdellovibrio bacteriovorus	X			X					
Xylella fastidiosa	X								
Chlamydophila pneumoniae	X								
Helicobacter hepaticus	X				X				
Legionella pneumophila	X			X		X			

Fig. 4. A fragment of the formal context for 20 bacteria.

It is evident directly from the context that these 20 bacteria constitute two clusters according to the Gram staining: there is no bacterium which is simultaneously Gram-positive and Gram-negative. Lattice diagrams on the Fig. 5 confirm this fact.

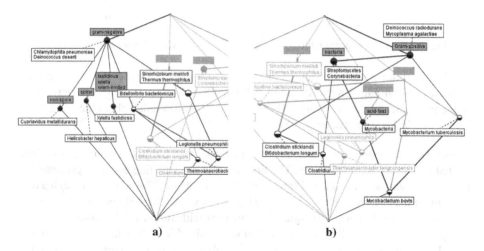

Fig. 5. Views of concept lattice demonstrating Gram staining: (a) – Gram-negative property, (b) – Gram-positive property.

Interpreting views on Fig. 5 as we did it for the example on Fig. 3 we resolve that bacteria are clustered according to their Gram staining because the views on Fig. 5(a) and (b) do not intersect.

Clustering bacteria according to the property of being aerobic is not evident from the context on Fig. 4. Lattice diagrams on Fig. 6 confirm the clustering bacteria according to this property in the same manner as for Fig. 4.

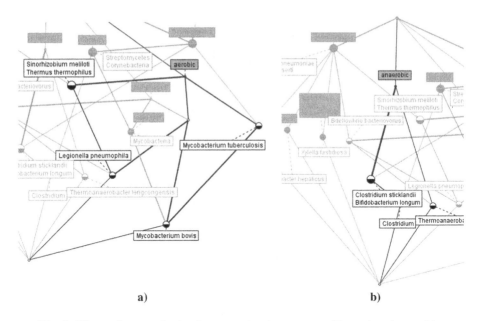

a)   b)

Fig. 6. Views of concept lattice demonstrating the property of bacteria to be aerobic.

However, the number of bacteria in Figs. 5 and 6 is not the same: Fig. 5 contains all 20 bacteria (10 in Fig. 5-a and 10 in Fig. 5-b.) and Fig. 4 - contains only 9 bacteria. This is due to the fact that the relevant texts do not contain information about the property of being aerobic for some bacteria.

Comparing Results. We can compare our results with two known similar solutions related to fact extraction problem. The first solution of extracting events is presented in [3] and is based on using special framework of EventMine. This solution is realized as marking of the text by highlighting its lexical elements as elements of event.

The second solution [24] is directly connected with BioNLP. The tasks of Named Entity Recognition and Relation extraction were solved in [24] with Alvis framework [27]. In [24] results of relations extraction are also presented as marked words in the texts. Our results of solution of NER are similar to [24] and presented in [6].

Comparing our current results of fact extraction with the known ones we resume that concept lattice provides principally another variant of solution of fact extraction problem. The main distinction of this solution is that it is not realized in the processed text by highlighting its lexical elements but it is realized with new external resource, conceptual model in the form of the concept lattice.

4   Conclusions and Future Work

This paper describes the idea of joining two paradigms of conceptual modeling - conceptual graphs and concept lattices. Current results of realizing this idea on textual data show its good potential for fact and knowledge extraction. Concept lattice may serve as a skeleton of ontology constructed on texts. Its data which may or may not be interpreted as facts constitutes a knowledge stored in the concept lattice being ready to extract.

In spite of the certain useful features of presented technology there are some problems which need to be solved for improving the quality of modeling technique.

1. Conceptual graphs acquired from texts contain many noisy elements. Noise is constituted by the text elements that contain no useful information or cannot be interpreted as facts. Noisy elements significantly decrease efficiency of algorithms of fact extraction.
2. Empirical rules which we use for creating formal contexts cannot embrace all configurations of conceptual graphs. More formal approach to creating formal contexts on the set of conceptual graphs will guarantee the completeness of solution. We guess that using pattern structures and their projections is the way of formalizing our modeling technique.
3. The next stage of developing current technology is creating of fledged information system which processes user queries and produces solutions of certain tasks on textual data. Not only visualization but also special user oriented interfaces to concept lattice will be created in this system.

Acknowledgments. The paper concerns with the work partially supported by the Russian Foundation for Basic Research, grant № 15-07-05507.

References

1. Aggarwal, C.C., Zhai, C.: Mining Text Data. Springer, New York (2012)
2. Hogenboom, F., Frasincar, F., Kaymak, U., de Jong, F., Caron, E.: A survey of event extraction methods from text for decision support systems. Decis. Support Syst. (DSS) **85**, 12–22 (2016)
3. Miwa, M., Ananiadou, S.: Adaptable, high recall, event extraction system with minimal configuration. BMC Bioinformatics **16**(10), 1–11 (2015)
4. BioNLP 2014. Workshop on Biomedical Natural Language Processing. Proceedings of the Workshop. The Association for Computational Linguistics, Baltimore (2014)
5. Bogatyrev, M., Tuhtin, V.: Creating conceptual graphs as elements of semantic texts labeling. In: Computational Linguistics and Intellectual Technologies, Proceedings of the International Conference "Dialogue", Moscow, pp. 31–37 (2009). (in Russian)
6. Bogatyrev, M.: Conceptual modeling with formal concept analysis on natural language texts. In: Proceedings of the XVIII International Conference "Data Analytics and Management in Data Intensive Domains", CEUR Workshop Proceeding, vol. 1752, pp. 16–23 (2016)
7. Bogatyrev, M., Samodurov, K.: Framework for conceptual modeling on natural language texts. In: Proceedings of International Workshop on Concept Discovery in Unstructured Data (CDUD 2016) at the Thirteenth International Conference on Concept Lattices and Their Applications. Moscow, CEUR Workshop Proceeding, vol. 1625, pp. 13–24 (2016)
8. Gildea, D., Jurafsky, D.: Automatic labeling of semantic roles. Comput. Linguist. **28**, 245–288 (2002)
9. Bogatyrev, M.Y., Vakurin, V.S.: Conceptual modeling in biomedical data research. Math. Biol. Bioinform. **8**(1), 340–349 (2013)

10. Bossy, R., Jourde, J., Manine, A.-P., Veber, P., Alphonse, E., Van De Guchte, M., Bessières, P., Nédellec, C.: BioNLP 2011 shared task - The bacteria track. BMC Bioinformatics, 13:S8, 1–15 (2012)

11. Campbell, N.A., et al.: Biology: Concepts and Connections. Benjamin-Cummings Publishing Company (2005)

12. Carpineto, C., Romano, G.: Exploiting the potential of concept lattices for information retrieval with CREDO. J. Univ. Comput. **10**(8), 985–1013 (2004)

13. Carpineto, C., Romano, G.: Using Concept Lattices for Text Retrieval and Mining. In: Ganter, B., Stumme, G., Wille, R. (eds.) Formal Concept Analysis. LNCS (LNAI), vol. 3626, pp. 161–179. Springer, Heidelberg (2005). doi:10.1007/11528784_9

14. Cimiano, P., Hotho, A., Staab, S.: Learning concept hierarchies from text corpora using formal concept analysis. J. Artif. Intell. Res. **24**, 305–339 (2005)

15. Edhlund, B., McDougall, A.: Pubmed Essentials, Mastering the World's Health Research Database. Form & Kunskap AB (2014)

16. Ganter, B., Kuznetsov, S.O.: Pattern structures and their projections. In: Delugach, H.S., Stumme, G. (eds.) ICCS-ConceptStruct 2001. LNCS (LNAI), vol. 2120, pp. 129–142. Springer, Heidelberg (2001). doi:10.1007/3-540-44583-8_10

17. Ganter, B., Stumme, G., Wille, R. (eds.): Formal Concept Analysis. LNCS (LNAI), vol. 3626. Springer, Heidelberg (2005). doi:10.1007/978-3-540-31881-1

18. Hunter, L., Cohen, K.B.: Biomedical language processing: what's beyond PubMed? Mol. Cell **21**, 589–594 (2006)

19. Ignatov, D.I., Kuznetsov, S.O., Poelmans, J.: Concept-based biclustering for internet advertisement. In: Vreeken, J., Ling, C., Zaki, M.J., Siebes, A., Yu, J.X., Goethals, B., Webb, G., Wu, X. (eds.) Proceeding of 12th IEEE International Conference on Data Mining Workshops (ICDMW 2012), pp. 123–130 (2012)

20. Meštrović, A.: Semantic matching using concept lattice. In: Proceeding Concept Discovery in Unstructured Data (CDUD-2012), pp. 49–58 (2012)

21. Otero, P.G., Lopes, G.P., Agustini, A.: Automatic acquisition of formal concepts from text. J. Lang. Technol. Comput. Linguist. **23**(1), 59–74 (2008)

22. Poelmans, J., Kuznetsov, S.O., Ignatov, D.I., Dedene, G.: Formal concept analysis in knowledge processing: a survey on models and techniques. Expert Syst. Appl. **40**(16), 6601–6623 (2013)

23. Priss, U.: Linguistic applications of formal concept analysis. In: Ganter, B., Stumme, G., Wille, R. (eds.) Formal Concept Analysis. LNCS (LNAI), vol. 3626, pp. 149–160. Springer, Heidelberg (2005). doi:10.1007/11528784_8

24. Ratkovic, Z., Golik, W., Warnier, P.: Event extraction of bacteria biotopes: a knowledge-intensive NLP-based approach. - BMC Bioinformatics, 13(Suppl 11):S8, pp. 1–11 (2012)

25. Sowa, J.F.: Conceptual Structures: Information Processing in Mind and Machine. Addison-Wesley, London (1984)

26. U.S. National Library of Medicine. http://www.ncbi.nlm.nih.gov/pubmed

27. Alvis framework. http://bibliome.jouy.inra.fr

28. Yevtushenko, S.A.: System of data analysis "Concept Explorer". In: Proceeding of the 7th National Conference on Artificial Intelligence KII-2000, Russia, pp. 127–134 (2000). (in Russian)

29. ConExp-NG. https://github.com/fcatools/conexp-ng

30. Wille, R.: Conceptual graphs and formal concept analysis. In: Lukose, D., Delugach, H., Keeler, M., Searle, L., Sowa, J. (eds.) Conceptual Structures: Fulfilling Peirce's Dream. ICCS-ConceptStruct 1997. LNCS (LNAI), vol. 1257, pp. 290–303. Springer, Heidelberg (1997). doi:10.1007/BFb0027878

Data Infrastructures in Astrophysics

Hybrid Distributed Computing Service Based on the DIRAC Interware

Victor Gergel[1], Vladimir Korenkov[2,3], Igor Pelevanyuk[2], Matvey Sapunov[4],
Andrei Tsaregorodtsev[3,5(✉)], and Petr Zrelov[2,3]

[1] Lobachevsky State University, Nizhni Novgorod, Russia
gergel@unn.ru
[2] Joint Institute for Nuclear Research, Dubna, Russia
{korenkov,pelevanyuk,zrelov}@jinr.ru
[3] Plekhanov Russian Economics University, Moscow, Russia
[4] Aix-Marseille University, Marseille, France
matvey.sapunov@univ-amu.fr
[5] CPPM, Aix-Marseille University, CNRS/IN2P3, Marseille, France
atsareg@in2p3.fr

Abstract. Scientific data intensive applications requiring simultaneous use of
large amounts of computing resources are becoming quite common. Properties
of applications coming from different scientific domains as well as their requirements to the computing resources are varying largely. Many scientific communities have access to different types of computing resources. Often their workflows
can benefit from a combination of High Throughput (HTC) and High Performance
(HPC) computing centers, cloud or volunteer computing power. However, all
these resources have different user interfaces and access mechanisms, which are
making their combined usage difficult for the users. This problem is addressed by
projects developing software for integration of various computing centers into a
single coherent infrastructure, the so-called interware. One of such software tool-
kits is the DIRAC interware. This product was very successful to solve problems
of large High Energy Physics experiments and was reworked to offer a general-
purpose solution suitable for other scientific domains. Services based on the
DIRAC interware are now proposed to users of several distributed computing.
One of these services is deployed at Joint Institute for Nuclear Research, Dubna.
It aims at integration of computing resources of several grid and supercomputer
centers as well as cloud providers. An overview of the DIRAC interware and its
use for creating and operating of a hybrid distributed computing system at JINR
is presented in this article.

Keywords: Grid computing · Hybrid distributed · Computing systems ·
Supercomputers · DIRAC

1 Introduction

Large High Energy Physics experiments, especially those running at the LHC collider
at CERN, have pioneered the era of very data intensive applications. The aggregated

L. Kalinichenko et al. (Eds.): DAMDID/RCDL 2016, CCIS 706, pp. 105–118, 2017.
DOI: 10.1007/978-3-319-57135-5_8

data volume of these experiments exceeds by today 100 PetaBytes, which includes both data acquired from the experimental setup as well as results of the detailed modeling of the detectors. Production and processing of these data required creation of a special distributed computing infrastructure - Worldwide LHC Computing Grid (WLCG). This is the first example of a large-scale grid system successfully used for a large scientific community. It includes more than 150 sites from more than 40 countries around the world. The sites altogether are providing unprecedented computing power and storage volumes. WLCG played a very important role in the success of the LHC experiments that achieved many spectacular scientific results like discovery of the Higgs boson, discovery of the pentaquark particle states, discovery of rare decays of B-mesons, and many others.

To create and operate the WLCG infrastructure, special software, so-called middleware, was developed to give uniform access to various sites providing computational and storage resources for the LHC experiments. Multiple services were deployed at the sites and centrally at CERN to ensure coherent work of the infrastructure, with comprehensive monitoring and accounting tools. All the communications between various services and clients are following strict security rules; users are grouped into virtual organizations with clear access rights to different services and with clear policies of usage of the common resources.

On top of the standard middleware that allowed building the common WLCG infrastructure, each LHC experiment, ATLAS, CMS, ALICE and LHCb, developed their own systems to manage their workflows and data and cover use cases not addressed by the middleware. Those systems have many similar solutions and design choices but are all developed independently, in different development environments and have different software architectures. This software is used to cope with large numbers of computational tasks and with large number of distributed file replicas by automation of recurrent tasks, automated data validation and recovery procedures. With time, the LHC experiments gained access also to other computing resources than WLCG. An important functionality provided by the experiments software layer is access to heterogeneous computing and storage resources provided by other grid systems, cloud systems and standalone large computing centers which are not incorporated in any distributed computing network. Therefore, this kind of software is often called interware as it interconnects users and various computing resources and allows for interoperability of otherwise heterogeneous computing clusters.

Nowadays, other scientific domains are quickly developing data intensive applications requiring enormous computing power. The experience and software tools accumulated by the LHC experiments can be very useful for these communities and can save a lot of time and effort. One of the experiment interware systems, the DIRAC project of the LHCb experiment, was reorganized to provide a general-purpose toolkit to build distributed computing systems for scientific applications with high data requirements [1]. All the experiment specific parts were separated into a number of extensions, while the core software libraries are providing components for the most common tasks: intensive workload and data management using distributed heterogeneous computing and storage resources. This allowed offering the DIRAC software to other user communities and now it is used in multiple large experiments in high energy physics, astrophysics and other domains. However, for relatively small user groups with little expertise in distributed computing, running dedicated DIRAC services is a very difficult task. Therefore, several computing infrastructure projects are offering DIRAC services as part of their services portfolio. In particular, these services are provided by the European Grid Infrastructure (EGI) project. This allowed many relatively small user communities to have an easy access to a vast amount of resources, which they would never have otherwise.

Similar systems originating from other LHC experiments, like BigPanDa [2] or AliEn [3] were also offered to use by other scientific collaborations. However, their usage is more limited than the one of DIRAC. BigPanDa is providing mostly the workload management functionality for the users and is not supporting data management operations, whereas DIRAC is a complete solution for both types of tasks. AliEn provides support for both data and workload management. However, it is difficult to extend for specific workflows of other communities. The DIRAC architecture and development framework is conceived to have excellent potential for extension of its functionality. Therefore, completeness of its base functions together with modular extendable architecture makes DIRAC a unique all-in-one solution suitable for many scientific applications.

This paper is an extension of the contribution to proceedings of the DAMDID'2016 Conference [15] providing a detailed description of the DIRAC service deployed in the Laboratory of Information Technologies in the Joint Institute for Nuclear Research (LIT/JINR), Dubna, Russia. The paper reviews the DIRAC Project giving details about its general architecture as well as about workload and data management capabilities in Sect. 2. Examples of the system usage are described in Sect. 3. Section 4 presents details of the DIRAC service deployment in LIT/JINR describing various components of the system as well as currently connected computing resources.

2 DIRAC Overview

DIRAC Project provides all the necessary components to create and maintain distributed computing systems. It forms a layer on top of third party computing infrastructures, which isolates users from the direct access to the computing resources and provides them with an abstract interface hiding the complexity of dealing with multiple heterogeneous services. This pattern is applied to both computing and storage resources. In both cases, abstract interfaces are defined and implementations for all the common computing service and storage technologies are provided. Therefore, users see only logical computing and storage elements, which simplifies dramatically their usage. In this section we will describe in more details the DIRAC systems for workload and data management.

2.1 Workload Management

The DIRAC Workload Management System is based on the concept of pilot jobs [4]. In this scheduling architecture (Fig. 1), the user tasks are submitted to the central Task

Queue service. At the same time the so-called pilot jobs are submitted to the computing resources by specialized components called Directors. Directors use the job scheduling mechanism suitable for their respective computing infrastructure: grid resource brokers or computing elements, batch system schedulers, cloud managers, etc. The pilot jobs start execution on the worker nodes, check the execution environment, collect the worker node characteristics and present them to the Matcher service. The Matcher service chooses the most appropriate user job waiting in the Task Queue and hands it over to the pilot for execution. Once the user task is executed and its outputs are delivered to the DIRAC central services, the pilot job can take another user task if the remaining time of the worker node reservation is sufficient.

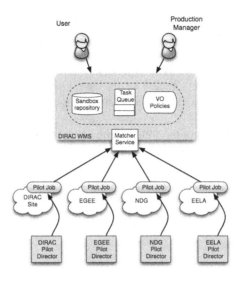

Fig. 1. WMS with pilot jobs

There are many advantages of the pilot job concept. The pilots are not only increasing the visible efficiency of the user jobs but also help managing heterogeneous computing resources presenting them to the central services in a uniform coherent way. Large user communities can benefit also from the ability of applying the community policies that are not easy, if at all possible, with the standard grid middleware. Furthermore executing several user tasks in the same pilot largely reduces the stress on the batch systems no matter if they are accessed directly or via grid mechanisms, especially if users subdivide their payload in many short tasks trying to reduce the response time.

The pilot job based scheduling system allows easy aggregation of computing resources of different technologies. Currently the following resources are available for DIRAC users:

- Computing grid infrastructures based on the gLite/EMI grid middleware. The submission is possible both through the gLite Workload Management System and directly to the computing element services exposing the CREAM interface. WLCG and EGI grids are examples of such grid infrastructures.

- Open Science Grid (OSG) infrastructure based on the VDT (Virtual Data Toolkit) suite of middleware [5].
- Grids based on the ARC middleware, which was developed in the framework of the Nordugrid project [6].
- Standalone computing clusters with common batch system schedulers, for example, PBS/Torque, Grid Engine, Condor, SLURM, OAR, and others. Those clusters can be accessed by configuring an SSH tunnel to be used by DIRAC directors to submit pilot jobs to the local batch systems. No specific services are needed on such sites to include them into a distributed computing infrastructure.
- Sites providing resources via most widely used cloud managers, for example Open-Stack, OpenNebula, Amazon and others. Both commercial and public clouds can be accessed through DIRAC.
- Volunteer resources provided with the help of BOINC software. There are several realizations of access to this kind of resources all based on the same pilot job framework.

As it was explained above, a new kind of computing resources can be integrated into the DIRAC Workload Management System by providing a corresponding Director using an appropriate job submission protocol. This is the plugin mechanism that enables easily new computing facilities as needed by the DIRAC users. Computing resources of some of those types are integrated within the JINR DIRAC service as described in Sect. 4.

2.2   Data Management

The DIRAC Data Management System (DMS) is based on similar design principles as the WMS [7]. An abstract interface is defined to describe access to a storage system with multiple implementations for various storage access protocols. Similarly, there is a concept of a FileCatalog service, which provides information about the physical locations of file copies. As for storage services there are several implementations for different catalog service technologies all following the same abstract interface.

A particular storage system can be accessible via different interfaces with different access protocols. But for the users it stays logically a single service providing access to the same physical storage space. To simplify access to this kind of services, DIRAC aggregates plugins for different access protocols according to the storage service configuration description. When accessing the service, the most appropriate plugin is chosen automatically according to the client environment, security requirements, closeness to the service, etc. As a result, users are only seeing logical entities without the need to know the exact type and technology of the external services.

DIRAC provides plug-ins for a number of storage access protocols most commonly used in the distributed storage services:

- SRM, XRootd, RFIO, etc.;
- gfal2 library based access protocols (DCAP, HTTP-based protocols, S3, WebDAV, etc.) [8].

New plug-ins can be easily added to accommodate new storage technologies as needed by user communities.

In addition, DIRAC provides its own implementation of a Storage Element service and the corresponding plug-in using the custom DIPS protocol. This is the protocol used to exchange data between the DIRAC components. The DIRAC StorageElement service allows exposing data stored on file servers with POSIX compliant file systems. This service helps to quickly incorporate data accumulated by scientific communities in any *ad hoc* way into any distributed system under the DIRAC interware control.

Similarly to Storage Elements, DIRAC provides access to file catalogs via client plug-ins. The only plug-in to an external catalog service is for the LCG File Catalog (LFC), which used to be a *de facto* standard catalog in the WLCG and other grid infrastructures. Other available catalog plug-ins are used to access the DIRAC File Catalog (DFC) service and other services that are written within the DIRAC framework and implement the same abstract File Catalog interface [9]. This plug-ins can be aggregated together so that all the connected catalogs are receiving the same messages on new data registration, file status changes, etc. The usefulness of aggregating several catalogs can be illustrated by an example of a user community that wants to migrate the contents of their LFC catalog to the DFC catalog. In this case, the catalog data can be present in both catalogs for the time of migration or for redundancy purpose.

The DIRAC File Catalog has the following main features:

- Standard file catalog functionality for storing file metadata, ACL information, checksums, etc.
- Complete Replica Catalog functionality to keep track of physical copies of the files.
- Additional file metadata to define ancestor-descendent relations between files often needed for applications with complex workflows.
- Efficient storage usage reports to allow implementation of community policies, user quotas, etc.
- Metadata Catalog functionality to define arbitrary user metadata on directory and file levels with efficient search engine.
- Support for dataset definition and operations.

The DFC implementation is optimized for efficient bulk queries where the information for large numbers of files is requested in case of massive data management operations. Altogether, the DFC provides logical name space for the data and, together with storage access plug-ins, makes data access as simple as in a distributed file system.

Storage Element and File Catalog services are used to perform all the basic operations with data. However, bulk data operations need special support so that they can be performed asynchronously without a need for a user to wait for the operation completion at the interactive prompt. DIRAC Request Management System (RMS) provides support for such asynchronous operations. Many data management tasks in large scientific communities are often repeated for different data sets. DIRAC provides support for automation of recurrent massive data operations driven by the data registration or file status change events.

In addition to the main DMS software stack, DIRAC provides several more services helping to perform particular data management tasks:

- Staging service to manage bringing data on-line into a disk cache in the SEs with tertiary storage architecture;
- Data Logging service to log all the operations on a predefined subset of data mostly for debugging purposes;
- Data Integrity service to record failures of the data management operations to spot malfunctioning components and resolve issues;
- The general DIRAC Accounting service is used to store the historical data of all the data transfers, success rates of the transfer operations.

2.3 DIRAC Development Framework

All the DIRAC components are written in a well-defined software framework with a clear architecture and development conventions. Since large part of the functionality is implemented as plug-ins implementing predefined abstract interfaces, extending DIRAC software to cover new cases is simplified by the design of the system. There are several core services to orchestrate the work of the whole DIRAC distributed system, the most important ones are the following:

- Configuration service used for discovery of the DIRAC components and providing a single source of configuration information;
- Monitoring service to follow the system load and activities;
- Accounting service to keep track of the resources consumption by different communities, groups and individual users;
- System Logging service to accumulate error reports in one place to allow quick reaction to occurring problem.

Modular architecture and the use of core services allow developers to easily write new extensions concentrating on their specific functionality and avoiding recurrent tasks.

All the communications between distributed DIRAC components are secure following the standards introduced by computational grids, which is extremely important in the distributed computing environment. A number of interfaces are provided to users to interact with the system. This includes a rich set of command-line tools for Unix environment, Python language API to write one's own scripts and applications, RESTful interface to help integration with third party applications. DIRAC functionality is available also through a flexible and secure Web Portal which follows the user interface paradigm of a desktop computer.

3 DIRAC Users

DIRAC Project was initiated by the LHCb experiment at CERN. LHCb stays the most active user of the DIRAC software. The experiment data production system ensures a constant flow of jobs of different kinds: reconstruction of events of proton-proton collisions in the LHC collider, modeling of the LHCb detector response to different kinds of events, final user analysis of the data [10]. The DIRAC Workload Management System ensures on average about 50 thousand jobs running simultaneously on more than 120 sites, with peak values going to up to 100 thousand jobs. This is equivalent to operating a virtual computing center of about 100 thousands of processor cores. LHCb exploits traditional grid computing resources, several standalone computing farms and also resources from several public cloud providers. An important special ingredient to the LHCb computing system is the High Level Trigger (HLT) farm, which is used for a fast "on-line" filtering of events during the normal data taking periods. In the periods when the LHC collider is stopped for maintenance operations, the HLT farm is used for modeling of the LHCb detector providing the total power of up to 35 thousands CPU cores. No particular batch system is used to manage those CPUs, which are under full control of the DIRAC WMS. At the same time the total data volume of LHCb exceeds 10 PB distributed over more than twenty millions of files, many of those having 2 and more physical copies in about 20 distributed storage systems. Information about all these data is stored in the DIRAC File Catalog. LHCb has created a large number of extensions to the core DIRAC functionality to support its specific workflows. All these extensions are implemented in the DIRAC development framework and can be released, deployed and maintained using standard DIRAC tools.

After the DIRAC system was successfully used within LHCb, several other experiments in High Energy Physics and other domains expressed interest in using this software for their data production systems, for example: BES III experiment at the BEPC collider in Beijing, China [11]; Belle II experiment at the KEK center, Tsukuba, Japan [12]; the CTA astrophysics experiment being constructed now in Chile [13], and others. Open architecture of the DIRAC project was easy to adapt for the workflows of particular experiments. All of them developed several extensions to accommodate their specific requirements all relying on the use of the common core DIRAC services.

3.1 DIRAC as a Service

Experience accumulated by running data intensive applications of the High Energy Physics experiments can be very valuable for researchers in other scientific domains, which have high computing requirements. However, if even the DIRAC client software is easy to install and use, running dedicated DIRAC services requires a high expertise level and is not easy especially for the research communities without long-term experience in large-scale computations. Therefore, several national computing infrastructure projects are offering now DIRAC services for their users. The first example of such service was created by the France-Grilles National Grid Initiative (NGI) project in France [14].

By 2011 in France, there were several DIRAC service installations used by different scientific or regional communities. There was also a DIRAC service maintained by the France-Grilles NGI as part of its training and dissemination program. This allowed several teams of experts in different universities to gain experience with installation and operation of DIRAC services. However, the combined maintenance effort for multiple DIRAC service instances was quite high. Therefore, it was proposed to integrate independent DIRAC installations into a single national service to optimize operational costs. The responsibilities of different partners of the project were distributed as follows. The

France-Grilles NGI (FG) ensures the overall coordination of the project. The IN2P3 Computing Centre (CC/IN2P3) hosts the service providing the necessary hardware and manpower. The service is operated by a distributed team of experts from several laboratories and universities participating to the project [15].

From the start, the FG-DIRAC service was conceived for usage by multiple user communities. Now it is intensively used by researchers in the domains of life sciences, biomedicine, complex system analysis, and many others. It is very important that user support and assistance in porting applications to the distributed computing environments is the integral part of the service. This is especially needed for research domains where the computing expertise is historically low. Therefore, the France-Grilles NGI organizes multiple tutorials for interested users based on the FG-DIRAC platform. The tutorials not only demonstrate basic services capabilities but are also used to examine cases of particular applications and the necessary steps to start running them on distributed computing resources. The service has an active user community built around it and provides a forum where researchers are sharing their experience and helping the newcomers.

After the successful demonstration of DIRAC services provided by the French national computing research infrastructure, similar services were deployed in some other countries: Spain, UK, China, and some others. There are several ongoing evaluation projects testing the DIRAC functionality and usability for similar purposes. Since 2014, DIRAC services are provided by the European Grid Initiative (EGI) for the research communities in Europe and beyond [16].

4 DIRAC in Joint Institute for Nuclear Research

A general-purpose DIRAC service was deployed in the Laboratory of Information Technologies (LIT) in JINR, Dubna. The Joint Institute for Nuclear Research (JINR) - is an international research center for nuclear sciences, with 5500 staff members, 1200 researchers including 1000 Ph.D's from eighteen member states countries. JINR participates in many experiments in physics domain all over the world. The DIRAC in JINR is provided to users participating in international collaborations that already use DIRAC tools. It is also used for evaluation of DIRAC as a distributed computing platform for experiments that are now under preparation in JINR. The service has several High Performance Computing (HPC) centers connected and offers a possibility to create complex workflows including massively parallel applications. The service is planned to become a central point for a federation of HPC centers in Russia and other countries. It will provide a framework for unified access to the HPC centers similar to existing grid infrastructures.

4.1 Computing Needs of JINR Experiments

The need to provide limited computing power and data storage occasionally appears from time to time. It could be at foundation of a new experiment like Nuclotron-based Ion Collider facility (NICA) [17], grows of existing one, like Baikal Deep Underwater

Neutrino Telescope (Baikal) [18] or decision of scientific group to move from computations on local PC's to more powerful dedicated computing resource. In JINR, there are several ways to provide necessary computing resources.

The first option is to use the batch system of the local computing center for calculations and storage. This is usually a good option for small groups and projects but it has the following constraints: access allowed only for user members in JINR, there is no guarantee that batch system will be able to process certain amount of tasks by certain time and command line interface which is the only way to work with batch could be a deterrent to some users.

The second option is a creation of a new dedicated computing infrastructure. This option could lead to big paper work, creation of a prototype and further upgrading in case of approval. Apart from that the one who is building such infrastructure should take into account that dedicated computing infrastructures tend to be idle considerable amounts of time and require many man power to maintain it. Usually only big scientific groups or laboratories could use this option.

DIRAC software was adopted in JINR as a viable solution to provide computing power and storage for different users. From the beginning the DIRAC service in JINR was designed to serve multiple experiments or user groups to reduce overall maintenance costs. Another point to use one DIRAC installation for different groups, or in terms of DIRAC Virtual Organizations, is the fact that under one installation it is possible and recommended to delegate management of all resource sharing policies to DIRAC and by doing that both reduce amount of work of a resource administrator and decrease the resource idle time.

Another big advantage of using DIRAC is the possibility to manage both storage and computing resources, which provides "single point of management" when it comes to pledges distribution and users and groups administration. Freed man-power could be invested either in development for particular Virtual Organization or in the development for the particular DIRAC installation which by itself could lead to the benefit of the whole DIRAC community. Also, using DIRAC across several groups could simplify user training and give a unified access to very different computing resources which users would never get in different circumstances like clouds, batch systems, grid computing elements or even supercomputers.

Despite the fact that administration work is reduced, operating the DIRAC service cannot be performed by any particular user group alone and should be handled centrally. In JINR, the Laboratory of Information Technologies is taking responsibility for maintaining and administrating the DIRAC service. The Laboratory of Information Technologies by itself has some amount of computing and storage resources which could be used for evaluation DIRAC by possible users and small groups but in case of serious requirements could be easily extended with additional external computing or storage capacity.

DIRAC is an open source software with the core written in Python and web part written in JavaScript. It is possible to modify, expand and tune DIRAC for the specific purposes in JINR. Since JINR participates in the BES III experiment [19], in particular contributing to construction of its computing system, a good expertise level in DIRAC [9] was developed, which helped to launch the general-purpose DIRAC service.

4.2   DIRAC Service in JINR

The DIRAC service at JINR is set up to aggregate various computing facilities in Dubna and at the sites of its partners in Russia and abroad. It is open for connection of centers of different types and at the point of writing includes five computing elements which are described in more details in the following.

The service is hosted on 3 virtual servers provided in the JINR cloud infrastructure:

dirac-conf.jinr.ru. This server for configuration and security services. Those services are security sensitive and require restricted access; therefore they are placed on a dedicated host. The following services are deployed on this server:

- The master Configuration Service is the primary database for all the DIRAC configuration data including description of computing resources, users and groups, community policies. This is also the primary tool for discovery of all the DIRAC distributed components;
- Several permanently running agents to synchronize the DIRAC Configuration Service with third party information sources, for example, information indices of grid computing infrastructures;
- User credential repository serves user identity information for the components that need to perform operation on behalf of particular users;
- Security logging service keeps track of all the authenticated connections to any DIRAC service to provide information necessary to investigate eventual security incidents;

dirac-services.jinr.ru. All the services supporting workload and data management are hosted on this server:

- Job scheduling and monitoring services include all the components to process user payloads from the initial reception up to delivering of the final results to users;
- Data catalogs to keep track of user file physical locations and ensure easy data discovery with user defined metadata;
- Services to support data operations like massive data replication and removal, data integrity checking and others;
- Computing resources monitoring and accounting services providing detailed information on the history of resources consumption

dirac-ui.jinr.ru. This server is hosting the DIRAC Web Portal and the RESTful API service. For the safety reasons, those components are separated from the core DIRAC services to reduce potential risks. In particular, these services do not have direct access to databases keeping the state of the whole system.

4.3   Computing Resources

The described services allow users to execute their payloads on various computing facilities which could be local in JINR or remote. One of the goals of the project is to collect in one system computing resources with different properties to demonstrate feasibility and advantages of providing a single user interface for all of them. In the following we present a brief description of these facilities.

DIRAC.JINR.ru represents the HybriLIT supercomputer cluster at the JINR Computing Center [20]. This cluster allows execution of parallel applications using MPI exchanges between the threads. It also supports calculations using accelerator co-processors, like GPGPU and others;

DIRAC.JINR-PBS.ru is a small cluster of commodity computers used by several experiments and groups in JINR;

DIRAC.JINR-SSH.ru is a collection of UNIX hosts not integrated in any computing cluster. The hosts are simply described in the DIRAC Configuration System with their IP addresses and SSH connection details. In this case, the DIRAC workload Management System plays the role of the batch system for this ad hoc computing cluster.

DIRAC.NNGU.ru represents the Lobachevsky Supercomputer of the University of Nizhny Novgorod [21]. A peculiarity of access to this facility is that it is done through a VPN tunnel in addition to the SSH secure access to the batch system gateway. Only a small subset of the supercomputer is accessible to the JINR DIRAC service, however, it demonstrates the possibility of remote connection with yet another set of access protocols;

DIRAC.AMU.fr is the supercomputer center of the Aix-Marseille University, France [22]. It supports parallel MPI applications and applications requiring the use of GPGPU's. Access is allowed to the local queues on the best effort basis when no other waiting priority tasks are available, which is very often the case and makes this center a valuable "opportunistic" resource.

More resources are planned for integration with the JINR DIRAC service to allow better possibilities for the users. In particular, resources provided by the JINR Openstack cloud will be an important use case for the purpose of demonstration of unified access for the users (Table 1).

Table 1. The computing facilities connected to the DIRAC service in JINR

Name	Access method	Batch system	Location	HPC support
DIRAC.JINR.ru	SSH	SLURM	JINR	yes
DIRAC.JINR-PBS.ru	SSH	PBS	JINR	no
DIRAC.JINR-SSH.ru	SSH	DIRAC	JINR + other	no
DIRAC.NNGU.ru	VPN + SSH	SLURM	Nizhny Novgorod, Russia	yes
DIRAC.AMU.fr	SSH	OAR	Aix-Marseille University, France	yes

5   Conclusions

DIRAC interware [23] is a versatile software suite for building distributed computing systems. It has gone a long way of development starting from a specific tool for a large-scale High Energy Physics experiment and is now available as a general-purpose product. Various computing and storage resources based on different technologies can be incorporated under the overall control by the DIRAC Workload and Data Management Systems. The open architecture of the DIRAC software allows easy connection of the new emerging types of resources as needed by the user communities. The system is designed for extensibility to support specific workflows and data requirements of particular applications. Completeness of its functionality as well as its modular design can ensure solution for a variety of distributed computing tasks and for a wide range of scientific communities in a single framework.

The DIRAC service in JINR is deployed with the goal to provide a platform for integration of heterogeneous computing resources with a simple and uniform interface for the users of experiments at JINR as well as for other user communities. It also plays the role of research and development facility to explore the possibility to integrate seemingly very different computing resources with different management systems and access methods. In particular, integration in one infrastructure of HTC and HPC computing resources helps to improve the efficiency of the resources usage and opens opportunities to create complex application workflows with different steps executed on different types of computers mostly suitable for the corresponding applications. Local and remote computing centers of different kinds are already connected to the service and demonstrated to work together with a single user interface. More resources, including clouds are planned for integration, which will provide more opportunities for the users of the service.

References

1. Tsaregorodtsev, A., Brook, N., Casajus Ramo, A., et al.: DIRAC3: the new generation of the LHCb grid software. J. Phys: Conf. Ser. **219**, 6 (2010). doi:10.1088/1742-6596/219/6/062029
2. Klimentov, A., Buncic, P., De, K., et al.: Next generation workload management system for big data on heterogeneous distributed computing. J. Phys: Conf. Ser. **608**, 1 (2015). doi:10.1088/1742-6596/608/1/012040
3. Bagnasco, S., Betev, L., Buncic, P., et al.: AliEn: ALICE environment on the grid. J. Phys: Conf. Ser. **119**, 6 (2008). doi:10.1088/1742-6596/119/6/062012
4. Casajus Ramo, A., Graciani, R., Tsaregorodtsev, A.: DIRAC pilot framework and the DIRAC workload management system. J. Phys: Conf. Ser. **219**, 6 (2010). doi:10.1088/1742-6596/219/6/062049
5. OpenScience Grid. https://www.opensciencegrid.org/
6. ARC project. http://www.nordugrid.org/arc/
7. Smith, A., Tsaregorodtsev, A.: DIRAC: data production management. J. Phys: Conf. Ser. **119**, 6 (2008). doi:10.1088/1742-6596/119/6/062046
8. Gfal2 Project. https://dmc.web.cern.ch/projects-tags/gfal-2
9. Poss, S., Tsaregorodtsev, A.: DIRAC file replica and metadata catalog. J. Phys: Conf. Ser. **396**, 3 (2012). doi:10.1088/1742-6596/396/3/032108
10. Stagni, F., Charpentier, P.: The LHCb DIRAC-based production and data management operations systems. J. Phys: Conf. Ser. **368**, 1 (2012). doi:10.1088/1742-6596/368/1/012010
11. Zhang, X.M., Pelevanyuk, I., Korenkov, V., et al.: Design and operation of the BES-III distributed computing system. Procedia Comput. Sci. **66**, 619–624 (2015). doi:10.1016/j.procs.2015.11.070
12. Kuhr, T.: Computing at Belle II. J. Phys: Conf. Ser. **396**, 3 (2015). doi:10.1088/1742-6596/396/3/032063

13. Arrabito, L., Barbier, C., Graciani Diaz, R., et al.: Application of the DIRAC framework in CTA: first evaluation. J. Phys: Conf. Ser. **396**, 3 (2015). doi: 10.1088/1742-6596/396/3/032007

14. France-Grilles DIRAC portal. http://dirac.france-grilles.in2p3.fr

15. Korenkov, V., Pelevanyuk, I., Tsaregorodtsev, A., Zrelov, P.: Accessing distributed computing resources by scientific communities using DIRAC services. In: The XVIII International Conference on Data Analytics & Management in Data Intensive Domains (DAMDID/RCDL 2016), Ershovo, Moscow Region, Russia, CEUR Workshop Proceedings, vol. 1752, pp. 110–115 (2016)

16. DIRAC4EGI service portal. http://dirac.egi.eu

17. Trubnikov, G., Agapov, N., Alexandrov, V., et al.: Project of the nuclotron-based ion collider facility (NICA) at JINR. In: EPAC 2008 Conference Proceedings, Genoa, Italy (2008)

18. Wischnewski, R.: The Baikal neutrino telescope – results and plans. Int. J. Modern Phys. A **20**(29), 6932–6936 (2005). doi:10.1142/S0217751X0503051X

19. Zweber, P.: Charm factories: present and future. AIP Conf. Proc. **1182**(1), 406–409 (2009)

20. Alexandrov, E., Belyakov, D., Matveyev, M., et al.: Research of acceleration of calculation in solving scientific problems on the heterogeneous cluster HybriLIT. RUDN J. Math. Inf. Sci. Phys. **4**, 20–27 (2015)

21. Lobachevsky supercomputer. https://www.top500.org/system/178472

22. Mesocentre of Aix-Marseille University. http://mesocentre.univ-amu.fr/en

23. DIRAC Project. http://diracgrid.org

Hierarchical Multiple Stellar Systems

Nikolay A. Skvortsov[1(✉)], Leonid A. Kalinichenko[1],
Dana A. Kovaleva[2], and Oleg Y. Malkov[2]

[1] Federal Research Center "Computer Science and Control" of Russian Academy of Sciences,
Moscow, Russia
nskv@mail.ru, leonidandk@gmail.com
[2] Institute of Astronomy of Russian Academy of Sciences, Moscow, Russia
{dana,malkov}@inasan.ru

Abstract. In astrophysics of hierarchical multiple stellar systems there is a contradiction between their maximum observed multiplicity (up to seven) and its theoretical limitations (about five hundred). To search for the hierarchical systems of high multiplicity we have analyzed modern catalogues of wide and close stellar pairs. We have compiled a list of objects - candidates for the stellar systems of maximum multiplicity, which includes an accurate cross-identification of their components. Presented procedure of cross-matching of multiple stellar systems is based on applying of criteria of unified form that exclude objects from sets of possible candidates to identification. Criteria are constructed using domain knowledge on astronomical objects of certain type. These criteria are not dependent on source catalogues but take into account knowledge on specific features of objects depending on conditions of observation.

Keywords: Binary stars · Visual binaries · Cross-matching · Conceptual modeling · Multi-component · Entity resolution

1 Introduction

The problem of cross-identification of celestial objects arises when working on virtually any tasks of astronomy, and is traditionally solved on a case-by-case basis. For single objects, this problem has been recognized and solved astronomical community since 80s of the last century. The problem of cross-identification of double and multiple stars is much more difficult. If a single star has, as a rule, only two celestial coordinates and magnitude, for a double star one needs to account magnitudes and coordinates of the reference and the concerned components and parameters of their orbital motion. This problem was discussed by astronomical community since the late 90s of the last century and has, in general, being solved by the authors when constructing Binary star database, BDB binaries data [11, 12]. By now BDB is the only resource of astronomical data that provides information about binary and multiple stars of all observational types. Finally, the solution for the problem of cross-identification of objects of higher multiplicity was developed for a number of special cases. In general, the difficulties arise because of a presence in the systems at the same time of the objects of different observational types: both wide, isolated (in evolutionary sense) pairs, and close pairs of stars, including

L. Kalinichenko et al. (Eds.): DAMDID/RCDL 2016, CCIS 706, pp. 119–129, 2017.
DOI: 10.1007/978-3-319-57135-5_9

eclipsing variables, X-ray sources and many others. Consequently, the number of parameters used for identification increases.

One of the objectives of high multiplicity stellar systems study is a search for hierarchical systems, confirming the theoretical justification for the possible existence of systems with a certain number of subordinate levels of stellar pairs. This issue was briefly described in [16] and is discussed in details in this paper.

Theoretical and observational multiplicity of hierarchical stellar systems is discussed in Sect. 2. Section 3 described catalogues of double stars used in the present study. In Sect. 4, principles of semiautomatic selection of high multiplicity systems are defined and results are presents. Efforts done to achieve automation and unification of cross-matching of systems and their components for future investigations of stellar systems with prospective catalogues are described in Sect. 5. Section 6 contain conclusions of the high multiplicity investigation.

2   Theoretical and Observational Stellar Multiplicity

Data on stellar multiplicity is important as a constraint on the problem of the formation and evolution of the Galactic stellar population. On the other hand, statistics of stellar multiplicity, i.e. the number of components, is poorly known, especially for multiplicities higher than three or four, and many questions still remain unresolved (see, e.g., the recent review in [4]).

The maximum number of components in a hierarchical multiple system depends on the number of hierarchy levels and can be estimated from theoretical considerations. A system is dynamically stable if, in a case of circular orbits, the outer orbital period exceeds the inner orbital period by a factor of five. For eccentric orbits, this factor is larger, increasing as $\sim (1-e)^3$ [21]. The mean outer/inner ratio for the semi-major axis and period is 20 and 70, respectively. On the other hand, the number of levels in hierarchical stellar systems is limited by the tidal action of regular gravitational field of the Galaxy, gravitational perturbations from passing stars, and stochastic encounters with giant molecular clouds (see, e.g., [8]). Surdin [18] demonstrated that, in these circumstances, the number of levels can reach values of 8 or 9, depending on masses and orbital parameters of the components. In the case of maximum dense "packing" of components in the system, hierarchical systems with 256 to 512 components can be produced.

On the other hand, there is no evidence to prove the existence of any hierarchical system having multiplicity of seven or higher. The most comprehensive catalogue of multiple systems [19] contains about 1350 hierarchical systems of multiplicity three to seven, and among the two catalogued septuple systems (AR Cas and ν Sco), at least the former one is a young cluster, i.e. is not necessarily hierarchical. This statistics is in a sharp contrast with the theoretical estimates given above. To eliminate this inconsistency, it is necessary to use additional sources of information, namely modern catalogues of double stars.

3   Catalogues of Double and Multiple Systems

Principal modern catalogues of visual double stars contain systems of much higher multiplicity than seven (see Table 1). Actually, WDS contains several systems of even higher multiplicity than indicated in Table 1. They represent either results of searches for sub-stellar companions to nearby stars by high-contrast and high-angular-resolution imaging, where at least some of the objects are background stars (WDS 17505−0603, 65 objects; WDS 19062−0453 = λ Aql, 107 objects) or results of speckle interferometric observations of stars in nebulae (WDS 05387−6906 = 30 Dor = Tarantula Nebula, 68 objects; WDS 05353−0523 = θ^1 Ori = Trapezium cluster in Orion nebula, 39 objects) or miniclusters/common proper motion groups (WDS 19147+1918, WDS 20315+3347, WDS 13447−6348, WDS 18354−3122, WDS 23061+6356, 38 to 44 objects per system).

Table 1. Principal catalogues of visual double and multiple systems. C, P, S are numbers of catalogued components, pairs, and systems, respectively

Catalogue, abbreviation, reference	C, P, S	M
The Washington Double Star Catalog (WDS [13])	249280, 133966, 115314	2–32
Catalogue of Components of Double and Multiple Stars (CCDM [3])	105837, 56513, 49325	1–18
Tycho Double Star Catalogue (TDSC [5])	103259, 37978, 64869	1–11

Note also in brackets that CCDM and TDSC contain some systems of multiplicity one. In the former case, this concerns astrometric binaries (with an invisible secondary component detected by its gravitational influence), while TDSC contains a fair amount of stars that the Tycho space mission failed to resolve into components.

It is instructive to plot the distribution of catalogued stellar systems on their multiplicity and compare it to the observational data. Tokovinin [20] presented statistics of catalogued multiple systems in the form N_i/N_{i-1} = 0.11, 0.22, 0.20, 0.36 for i = 3, 4, 5, 6, respectively, where N_i is the number of systems of the i-th multiplicity. These results are in a good accordance with conclusions made in their study [14] of multiple objects in the immediate (closer than 25 pc) solar vicinity. Later Tokovinin [22] studied hierarchical systems among F and G dwarfs in the Solar neighborhood and found the fraction of multiple systems with 1, 2, 3,… components to be 54:33:8:4:1. Our comparison shows that the most complete catalogue, WDS, satisfactory images the [14, 20] distributions, while the newer and deeper study [22] demonstrates a surplus of triple and higher multiplicity systems in comparison with the catalogued systems (or, conversely, WDS contains superfluous, obviously optical, double stars).

The listed catalogues of visual binaries contain various data for evidently overlapping sets of objects, and no one of them contains all known visual systems. Thus, to use the complete dataset, it was necessary to cross-match these catalogues, i.e. to gather all available information on visual binary stars in a single list. A comprehensive set of visual binaries using data from the current versions of the three listed catalogues was compiled in [7].

4 Selection of High Multiplicity Systems

The applied in [7] cross-matching procedure worked quite well for systems of multiplicity about five or six, but often failed to correctly cross-identify components in systems of higher multiplicity, due to high spatial density of objects.

To compile a list of candidates to hierarchical stellar systems of maximum multiplicity (and estimate the value of this maximum multiplicity), as well as to finally solve the problem of cross-identification of multiple systems, we have performed a semi-automatic identification of systems of multiplicity six and more in principal catalogues of visual double and multiple systems (see Table 1). The total number of such systems is 551. 175 of them are included in WDS only. The remaining 395 systems are included in more than one catalogue and, consequently, their components need cross-matching (the systems themselves were cross-matched in [7] and analyzed in [10].

Compiling the list of very high multiplicity systems, we were flagging optical pairs. The information about non-physical nature of a pair can be found in WDS and the textual Notes to WDS. We have also applied the criterion to select optical pairs suggested in [15], which revealed additional optical objects. Optical pair selection is described in detail in [6].

Photometrically unresolved binarity of some components can increase actual multiplicity of a system. In order to take this into account, we have cross-matched our systems with lists of closer binaries (orbital, interferometric, spectroscopic, eclipsing, X-ray systems, radio pulsars) using the Binary star database, BDB [9, 11]. Besides, indication of hidden binarity can sometimes be found in WDS Notes (34 cases). Information on sub-components of our very high multiplicity systems was found in catalogues of orbital (77 pairs), interferometric (425 pairs), spectroscopic (52), and eclipsing (16) binaries.

Table 2. Multiple system statistics. M: multiplicity of systems; N1: number of candidate systems; N2: number of prospective systems; N3: number of confirmed systems.

M	N1	N2	N3	M	N1	N2	N3	M	N1	N2	N3
6	138	54	5	15	14	1	–	24	4	–	–
7	107	34	3	16	4	3	–	25	2	1	–
8	76	20	–	17	5	2	–	26	1	–	–
9	42	16	–	18	6	2	–	27	1	–	–
10	39	13	1	19	4	–	–	28	–	1	–
11	32	9	–	20	3	–	–	29	–	1	–
12	23	3	1	21	2	1	–	30	1	1	–
13	14	4	–	22	3	–	–	31	2	1	–
14	13	1	–	23	–	–	–	32	1	–	–

Finally, we have excluded from our statistics those pairs that have no clear indication on their physical binarity, according to WDS and the textual Notes to WDS. As a result, we have compiled a list of 10 so-called "confirmed" systems of multiplicity six and higher. The list contains all systems included in the WDS, CCDM, and TDSC catalogues, and thus it is the most comprehensive list of stellar systems of multiplicity six and more. We provide extensive cross-identifications for components, pairs, and systems included in the list. We add data on photometrically unresolved binaries, taken from catalogues of closer pairs (spectroscopic, eclipsing, etc.) and flag optical pairs.

The final statistics is presented in Table 2. Column N1 contains the number of candidates to systems of multiplicity M Column N2 contains the number of candidates to systems of multiplicity M, without optical pairs Column N3 contains the number of confirmed systems of multiplicity M.

The highest-multiplicity systems are listed in Table 3. For each system, the number of components (M1), number of optical components (Opt), and number of confidently hierarchical components (M2) are given. The system WDS 17457−2900 demonstrates the highest value of possible hierarchical multiplicity (7), while possible multiplicity of several other systems (WDS 23061+6356, WDS 17378−1315, WDS 10174−5354) reaches higher values, but it should be confirmed by observations.

Table 3. Systems of highest prospective multiplicity. M1: number of components without optical ones; Opt: number of optical pairs; M2: possibly confident hierarchical multiplicity.

ID	M1	Opt	M2
WDS 23061+6356	31	8	1
WDS 17378−1315	30	2	1
WDS 10174−5354	29	3	2
WDS 10451−5941	28	6	1
WDS 15326−5221	25	0	1
WDS 17457−2900	21	0	7
WDS 01030+6914	18	1	1
WDS 05353−0522	18	0	1

It can be seen that these values are still far from those expected from theoretical predictions. Several possible ways can be considered to explain such a mismatch.

First, the theoretical possibility to construct a system with 8–9 hierarchy levels is based on purely geometrical considerations and does not necessarily mean that physical conditions in a protostellar cloud can permit to construct such a system. Consecutive fragmentation of a large contracting interstellar cloud is needed for a very high multiplicity hierarchical system to be born.

Also, very wide binaries (wider than 100 AU) are so weakly bound that they can be effectively disturbed, even disrupted, by extremely weak perturbations from inhomogeneities in the Galactic potential due to stars, molecular clouds, dark objects, or large-scale tides. Thus, the outermost components of a very high multiplicity hierarchical system will probably not survive on their orbits and leave the system.

Finally, we probably underestimate hidden multiplicity of stellar systems, and the number of photometrically unresolved components is much higher than catalogued data predict.

5   Procedure of the Multiple Systems Cross-Matching

The results of semiautomatic selection of stellar systems are useful for further advancing of cross-matching algorithms. Automation of multiple stellar systems remains an ongoing issue. The methodology of multiple entity matching proposed below is intended to be applied to systems of arbitrary multiplicity, its results are not dependent on the sequence of component identification. It may be useful for any existing and prospective catalogues since it is not configured for particular catalogues, but focused on consideration of all available domain knowledge including behavior of observational and astrophysical characteristics of given type of objects as well as characteristics of observation conditions.

Data from various astronomical catalogues of single and multiple stars of different observational types should be analyzed to identify components of stellar systems through all sources. The problem of stellar systems identification is reduced to matching of multicomponent entities from multiple data sources. Components of such entities may be typed and characterized in different data sources by different sets of attributes. A component may represent a multiple object in its turn in some data sources. For example, a close pair of stars may be discovered with spectral methods but it is catalogued as a single star in visual observations.

A multiple system is represented as a graph with components (or stellar objects unresolved yet to be multiple) as vertices, and pairs of components from catalogues denoted by the arcs from primary component to secondary one in a pair. Identification of the whole stellar systems is equivalent to construction of the connected graphs from the identified components. Every vertex, arc and graph as a whole system should be correctly identified. Any wrong identification of components or pairs obviously may cause a wrong connection of several systems into a single one, or single stars into the systems.

Matching of star systems involves a set of entity resolution approaches. Approaches to matching include various criteria based on sets of attributes, graph structures and identifiers of stars. Matching methods use not only data of observations and parameters of objects, but also identification based on objects that already have been identified [1, 2].

During analysis of data, a number of issues should be taken into account:

- different formatting data in catalogues;
- different semantics of attributes in catalogues (e.g. in different catalogues stellar coordinate may refer to the photocenter of the pair or to a brighter component in the pair);
- data input errors in catalogues (for instance, errors in the identifiers of stars);
- missing values in catalogue fields;
- variable values of attributes (for instances, changing brightness, changing coordinates in different observations as a result of proper motion or orbital movement of components);
- unstructured data (comments useful for identification of components);
- differences in representation of system structure (for example, different components may be deemed main in the pair in case of the similarity of their characteristics);

- depending on angular distance of components, components' brightness difference and resolution of a catalogue a pair may be catalogued as a single object with integral brightness or as two separate objects;

Catalogues usually contain some fields with identifiers of objects from other known catalogues. However, objects in different catalogues may have implicitly different semantics. Therefore, every identification taken from original catalogues as a cross-catalogue identifier may be wrong for some reasons and needs to be verified in a possible way, for instance, using values of observed parameters and calculated astrophysical parameters of identified entities.

5.1 Conceptualization of the Domain

Specifications of subject domain include all the concepts and knowledge used in matching of multiple systems. These include a quality of the observation instruments (angular resolution, lower and upper magnitude limits), astrometry (coordinate systems, proper motion, orbital movement, separation and position angle), photometry (photometric systems, passbands, magnitudes, variability), pair classification according to the observational technique (spectroscopic, interferometric, eclipsing pairs). Using ontological concepts, the conceptual schema is constructed for unified representation of data from heterogeneous sources [17]. Concept constraints are used for generation of sets of criteria for entity matching.

5.2 Technical Approach to Matching

General methodology for single or multiple entity matching is based on construction of a set of candidates for every entity of its component and application of sets of criteria constraining such sets of candidates. Criteria are formed from the domain knowledge limiting the interpretation of objects: expression of secondary attributes (computable) from primary ones (observed parameters), constraints of attribute values, mutual constraints of two or more attributes. Criteria have equal importance, take entities, not parameters as arguments, and are applied when all required data on entities is present. Order of application of the criteria has influence only to the effectiveness of candidate set constraining. For this purpose, the criteria may have a priority of application.

The approaches used for the entity resolution include various similarity criteria for subsets of attribute values (based on domain knowledge) and graph structures, which allow estimating identity of systems and components. Matching of entities may also include identification criteria based on the identifications of entities that already established.

It is reasonable to divide the process of entity matching into several interacting stages of matching candidate search. Firstly, components of systems from surveys of visual binaries are cross-matched as single objects. Then on the base of the first stage results, visual pairs are cross-matched (as the widest pairs in the hierarchical systems). On the next stage, consideration is complemented with the pairs of the other observational types (close pairs). Finally, existing identifiers of systems, pairs and components are cross-matched using results of the previous stages. Matching of the whole systems is a consequence of matching of all their components and pairs.

In the following subsections we describe the mentioned stages in more details.

5.3 Component Matching

For every component a set of candidates for identification in all considered catalogues (including surveys not differentiating single and multi-component objects) is constructed. Construction of candidate sets of component identity is primarily based on the proximity of the coordinate values. These sets then are used in other criteria verifying combination of parameter values such as:

- taking into account effective angular resolution, trigonometric parallax and difference due to proper motion and observational epochs;
- similar brightness in a given photometric passband taking into account sensibility to lower and higher limits of magnitude;
- limiting of brightness or color index difference in different photometric passband;
- taking into account possible variability of the star;
- similarity of evolution statuses;
- similarity of spectral classification.

In addition, we use special criteria to return from the next stages and to apply their results to choose correct candidates for component identification.

5.4 Visual Pair Matching

Work on identifying pairs of components begins with the consideration of wide visual ones. For every pair a set of candidates for identification includes possible combinations of identified components from results of the previous stage. For each candidate for pair identification, positional and photometric information is compared in criteria such as:

- position angle and separation of secondary component to primary one may differ because of the orbital movement;
- significant difference of the component proper motions in case of optical pairs (located close in the sky, but far apart in space);
- proper motions of components in a pair are similar;
- differences of component brightness are similar for identical pairs;
- sometimes it is possible to compare chemical compositions and evolution statuses of components in physical pairs;
- unique identification of components and other pairs are taken into account.

Criteria are constructed using the domain knowledge and results of statistical analysis of data from different catalogues or initial source surveys of subsets in catalogues. Limit values of parameter deviations are determined. If difference of parameter values does not exceed limiting value of the parameter, it may be a criterion of the pair identification.

Sometimes a pair should be matched not to another pair but to a component. A pair of the nearby stars depending on brightness and angular distance may be catalogued by instruments with different angular resolution as a single object (having integral brightness) or as the two distinct objects. To determine such cases an effective angular resolution of a catalogue is calculated statistically.

There are several methods to detect optical pairs. A sign of optical pair may be a difference of proper motions of components and/or their annual parallax. If there are series of observations, another sign is a linear (not orbital) relative movement of components. One more known statistical method for optical pair detection is the so-called 1% filter method [15] using field density in the direction of galactic coordinates of components, brightness of secondary components and angular distance between components. The identified probable optical pairs are marked with a special flag.

5.5 Close Pair Matching

Results of visual pair matching facilitates involvement of the close pairs. At this stage, data on binary and multiple systems of the following observational types is considered: interferometric, orbital, astrometric, spectroscopic, eclipsing, cataclysmic, X-ray binaries, as well as binaries of radio pulsars. Such pairs may be autonomous systems or components of previously considered wide visual pairs. Sets of matching criteria are mostly based on positional and photometric information but have some special criteria for specific parameters of objects. It should be noted that a pair may be listed in different catalogues as objects of different observational types.

5.6 Identifier Matching

At final stage, results of component and pair matching are complemented with identifiers from original catalogues of multiple and single stars (Bayer/Flamsteed, DM, HD, GCVS, HIP). These identifiers are commonly and widely used. However, the problem of belonging of these identifiers to a pair or to one (or the other) component in that pair often requires careful consideration. Correct identification of pairs and components using all available observed parameters on previous stages helps to solve this problem. Belonging of an identifier to a pair or to a component depends also on the difference of effective resolutions of original catalogue.

There are criteria that detect different types of identification errors. For example, an assumption of mixed up identification of components in a pair is generated if differences of brightness in different catalogues have close absolute value, but with different sign.

For each system, pair and component a specific identifier is assigned which is associated with the identifiers of all original catalogues of single and multiple stars to form a common base of identifications. Objects not resolved automatically are signed with special flags and are considered by experts for manual resolution or criterion base modification.

6 Conclusions

To explain inconsistency in stellar multiplicity between rather high values predicted by theoretical considerations and observational lack of systems with multiplicity higher than six, we have studied principal catalogues of visual double stars: WDS, CCDM and TDSC. They contain data on very high multiplicity (up to 30 components and more), though not necessarily hierarchical, systems, including moving groups and (mini-)clusters. To collect all available information on these systems, it was first necessary to make a thorough and accurate cross-matching of their components in the catalogues. Optical pairs, when known or assumed from the probability filter, were flagged and eliminated from the statistics, and information on photometrically unresolved sub-components was added.

Principal results of the current study are the following:

- a cross-identification catalogue of 551 stellar systems of multiplicity six and more;
- a list of systems, candidates to utmost multiple hierarchical systems;
- a procedure for cross-matching components of very high multiplicity systems (i.e., in crowded stellar fields), which also can be used for identification of objects in future surveys of binary/multiple stars (Gaia, LSST).

Acknowledgments. The work was partly supported by the Presidium of the Russian Academy of Sciences program "Leading Scientific Schools Support" 9951.2016.2 and by the Russian Foundation for Basic Research grants 15-02-04053, 16-07-01028, and 16-07-01162.

This research has made use of the VizieR catalogue access tool (the original description of the VizieR service was published in A&AS 143, 23), the SIMBAD database, operated at the Centre de Donnees astronomiques de Strasbourg, the Washington Double Star Catalog maintained at the U.S. Naval Observatory, and NASA's Astrophysics Data System Bibliographic Services.

References

1. Bhattacharya, I., Getoor, L.: Entity resolution in graphs. In: Cook, D.J., Holder, L.B. (eds.) Mining Graph Data, pp. 311–332. Wiley, USA (2006)
2. Christen, P.: Data Matching: Concepts and Techniques for Record Linkage, Entity Resolution, and Duplicate Detection, p. XX, 272. Springer Science & Business Media, Heidelberg (2012). ISBN: 978-3-642-31164-2
3. Dommanget, J., Nys, O.: VizieR on-line data catalog: I/274 (2002). http://cdsarc.u-strasbg.fr/viz-bin/Cat?I/274
4. Duchene, G., Kraus, A.: Stellar multiplicity. Ann. Rev. Astron. Astrophys. **51**, 269–310 (2013)
5. Fabricius, C., Hog, E., Makarov, V.V., et al.: The Tycho double star catalogue. A&A **384**, 180–189 (2002)
6. Gebrehiwot, Y.M., et al.: On utmost multiplicity of hierarchical stellar systems. Baltic Astron. **25**, 393–399 (2016)
7. Isaeva, A.A., Kovaleva, D.A., Malkov, O.Y.: Visual binaries: cross-matching and compiling of a comprehensive list. Baltic Astron. **24**, 157–165 (2015)
8. Jiang, Y.-F., Tremaine, S.: The evolution of wide binary stars. MNRAS **401**, 977–994 (2010)
9. Kaygorodov, P., Debray, B., Kolesnikov, N., et al.: The new version of Binary star database (BDB). Baltic Astron. **21**, 309–318 (2012)

10. Kovaleva, D.A., Malkov, O.Y., Yungelson, L.R., et al.: Statistical analysis of the comprehensive list of visual binaries. Baltic Astron. **24**, 367–378 (2015)
11. Kovaleva, D.A., et al.: Binary star DataBase BDB development: structure, algorithms, and VO standards implementation. Astron. Comput. **11**, 119–125 (2015)
12. Malkov, et al.: Binary star database BDB: datasets and services. Astron. Astrophys. Trans. **28**, 235–244 (2013)
13. Mason, B.D., et al.: VizieR on-line data catalog: B/wds (2016). http://cdsarc.u-strasbg.fr/viz-bin/Cat?B/wds
14. Orlov, V.V., Titov, O.A.: Statistics of nearby star population. Astron. Rep. **38**, 462–467 (1994)
15. Poveda, A., Allen, C., Parrao, L.: Statistical studies of visual double and multiple stars. I - Incompleteness of the IDS, intrinsic fraction of visual doubles and multiples, and number of optical systems. Astrophys. J. **258**, 589–604 (1982)
16. Skvortsov, N.A., Kalinichenko, L.A., Kovaleva, D.A., Malkov, O.Y.: Search for hierarchical stellar systems of maximal multiplicity. In: Selected Papers of the XVIII International Conference on Data Analytics and Management in Data Intensive Domains, vol. 1752, pp. 219–225. CEUR Workshop Proceedings (2016)
17. Skvortsov, N.A., et al.: Conceptual approach to astronomical problems. Astrophys. Bull. **71**(1), 114–124 (2016). Springer
18. Surdin, V.G.: Hierarchical star clusters: fractal properties and maximum population. ASP Conf. Ser. **228**, 568–570 (2001)
19. Tokovinin, A.A.: MSC – A catalogue of physical multiple stars. A&AS **124**, 75–84 (1997)
20. Tokovinin, A.A.: Statistics of multiple stars: some clues to formation mechanisms. In: Zinnecker, H., Mathieu, R.D. (eds.) Proceedings of IAU Symposium 2001, The Formation of Binary Stars, Potsdam, pp. 84–92. ASP, San Francisco (2001)
21. Tokovinin, A.A.: Statistics of multiple stars. In: Allen, C., Scarfe, C. (eds.) Revista Mexicana de Astronomía y Astrofísica Serie de Conferencias, vol. 21, pp. 7–14. Instituto de Astronomia, UNAM, Mexico (2004)
22. Tokovinin, A.A.: From binaries to multiples. II. Hierarchical multiplicity of F and G Dwarfs. Astron. J. **147**(4), 87–101 (2014)

Observations of Transient Phenomena in BSA Radio Survey at 110 MHz

Vladimir A. Samodurov[1,2(✉)], Alexey S. Pozanenko[3,4],
Alexander E. Rodin[1], Dmitry D. Churakov[5], Dmitry V. Dumskij[1,2],
Evgeny A. Isaev[1,2], Andrey N. Kazantsev[1], Sergey V. Logvinenko[1],
Vasily V. Oreshko[1], Maxim O. Toropov[6], and Maria I. Volobueva[7]

[1] Pushchino Radio Astronomy Observatory ASC LPI, Pushchino, Russia
{sam,rodin,dumsky,lsv,oreshko}@prao.ru,
eisaev@hse.ru, kaz.prao@bk.ru
[2] National Research University Higher School of Economics, Moscow, Russia
[3] Space Research Institute, Moscow, Russia
apozanen@iki.rssi.ru
[4] Moscow Engineering Physics Institute,
National Research Nuclear University MEPhI, Moscow, Russia
[5] TsNIIMash, Korolev, Russia
dmitr22@list.ru
[6] OOO "Business Automation", Moscow, Russia
mtl710@yandex.ru
[7] Saint Petersburg State University, St. Petersburg, Russia
panther_gatchina@mail.ru

Abstract. One of the most sensitive radio telescopes at the frequency of 110 MHz is a Big Scanning Antenna (BSA) in Pushchino Radio Astronomy Observatory of Lebedev Physical Institute (PRAO LPI, Moscow region, Russia). Since 2012 in the BSA the continuous survey observation was started in multibeam mode in the frequency band of 109–112 MHz. Now 96 beams covering from −8 and up to +42° in declination are used. The number of frequency bands are 6 with a time resolution of 0.1 s and 32 bands with the time resolution of 0.0125 s. In a fast mode (32 bands, 0.0125 s) daily data flow is 87.5 GB (32 TB per year). The data provide a great opportunity for both short-term and long-term monitoring of the various radio sources. The sources are fast radio transients of different nature, such as fast radio bursts (FRB), possible counterparts of gamma-ray bursts (GRB), and sources of gravitational waves, the Earth's ionosphere, interplanetary and interstellar plasma. Based on the BSA observations the database is constructed. We discuss data base properties, the methods of transient search and allocation in database. Using this database we were able to detect 83096 individual transient events in the period of July 2012 – October 2013, which may correspond to pulsars, scintillating sources and fast radio transients. We also present first results and statistics of transients classification. In particular we report parameters of two candidates in new RRAT pulsars.

Keywords: Radioastronomy · Survey · Pulsars · Fast radio bursts · Transients

© Springer International Publishing AG 2017
L. Kalinichenko et al. (Eds.): DAMDID/RCDL 2016, CCIS 706, pp. 130–141, 2017.
DOI: 10.1007/978-3-319-57135-5_10

1 Introduction

One of the most sensitive radio telescopes at the frequency of 110 MHz is a Big Scanning Antenna (BSA), also known as the Big Cophasal Antenna. BSA LPI radio telescope was constructed in 1970–1974 by Lebedev Physical Institute of the Russian Academy of Sciences. BSA is the radio telescope of meridian type. It is constructed as equidistant phased array antenna consisting of 16384 wave dipoles located on a square of 384 × 187 m (geometric area of BSA is over 70 thousand square meters, and the effective area is about 45 thousand square meters).

Radio telescope originally operated in a frequency band of 101–104 MHz, was reconfigured in 1996 to the band of 109–112 MHz (the wavelength of about 3 m). In this range it is the most sensitive telescope of meridian type in the world. The system equivalent flux density (SEFD) of the telescope is 34 Jan [1] at the zenith and at the minimum temperature of the background, which is about 3 times better than that of the radio telescope LOFAR [2] at the frequency of 110 MHz.

BSA LPI is a unique tool for multiple investigations of pulsed sources and transients of different nature: pulsars, study of dynamic processes in the near-solar and interplanetary plasma, the compact radio source structure analysis in the meter wavelength. Nowadays a long wavelength radio astronomy (in the range of 10 to 300 MHz) is experiencing a new period of interest. There is a wider class of scientific problems for this frequency range (see, e.g., [3]). Existing telescopes (the most famous of them is UTR-2 [4]) are refurbishing, and new radio telescopes are under construction. Besides LOFAR, the GURT [3] and LWA [5] should be noted.

The paper is based on the enlarged version of [11]. In particular, we have added statistics of transient events detected by BSA, report for the first time details of discovered candidates in Rotating Radio Transients (RRAT) and discuss future plans for access to a BSA data. The paper is organized as follows. In the Sects. 1.1–1.2 we provide a brief description of equipment and data acquisition. The Sect. 2 describes the data processing and a database. In Sect. 3 we present a preliminary classification of found transients. The Sect. 4 discusses the data dissemination.

1.1 Specifications of the BSA LPI Radio Telescope

The most important feature of the BSA is an ability of receiving in a full power mode. This mode allows to observe not only point-like sources but also a background radiation of our Galaxy and extended sources with a specific size up to 2–3°. The second feature is multibeam mode of operation. Before 2007 the BSA was designed to generate 16 beams, followed by a second, independent 16-beam diagram formed. Finally, in 2010–2011 (the third) diagram beam forming system of 128 beams was designed and constructed.

The main technical parameters of the BSA radio telescope are the following:

- Operating frequency band 109–112 MHz;
- The effective area of the antenna (maximum) - 47 000 m^2;
- The system noise temperature (minimum) - 560 K;

– Antenna polarization - linear (horizontal, along the east-west direction);
– The number of simultaneously generated beams - 128;
– The width of beam diagram in the E-plane - 54 arcmin;
– The width of beam diagram in the H-plane - 24 arcmin (zenith).

A new system allows to form 128 beams of the BSA diagram which covers the field of view of 40 square degrees. The first 48 beams of the diagram were equipped with multi-channel receivers on July 7, 2012 (the start of continuous survey). Since April 1, 2013 the observational data is recording with 96 beams in the sector of declinations of $-8° < δ < +42°$ toward South.

Characteristics of receiving and recording system (for ideal conditions: sensitivity - for the zenith, and at the minimum background temperature):

– the number of beams – 96 (32 beams out of 128 currently are not used);
– band registration frequencies - 2.5 MHz, the center frequency - 110.25 MHz;
– number of frequency channels per beam - 6 to 32;
– signal sampling interval in the frequency channel $Δt = 12.5, 100$ ms;
– maximum sensitivity (the entire band of 2.5 MHz) with a sampling of 100 ms - 0.07 Jy, (12.5 ms - 0.2 Jy);
– the sensitivity for a spectral band (415 kHz or 78 kHz, respectively): 0.16 or 1.06 Jy for 100 and 12.5 ms, respectively.

Thus, starting from April 1, 2013, the BSA is recording data obtained with 96 beams on the South in the declination range $-8° < δ < +42°$ (since July 6, 2012 registration was recording with 48 beams, in the declination range $+8.7° < δ < +12.6°$ and $25° < δ < +42°$). The observations covers 5.08 steradian per day or 0.40 surface of the whole sphere. Multibeam diagram simultaneously covers 40 square degrees ($\sim 1/1000$ entire sphere).

1.2   Registration of the Observational Data

As a data recorder an industrial computer is used with the possibility of installing a set of 6 modules for recording signals coming from different beams of the BSA radio telescope. Each module processes and records data of 8 beams. As a result, one recorder is registering data from 48 beams. Totally two recorders are in operation now.

The signal from each beam is analyzed and digitized in several frequency bands. The standard mode is of 6 bands for a full bandwidth of 2.5 MHz. In addition, a signal digitized from full bandwidth of 2.5 MHz is also recorded. For a given number of 6 bands (the bandwidth of each is 415 kHz), the output file will contain information about the 7 frequency bandwidths. When the number of bands 32 (the bandwidths of each band is 78 kHz), the number of recording channels is 33. The data are digitized with time sampling 100 ms and 12.5 ms. The digitized data consist of 32-bit floating-point numbers. In the standard mode (6 bands on 415 kHz, time sampling 100 ms) an array of four-byte numbers of 48×7 is recorded ten times per second. At the end of every hour the accumulated data are recorded in a file (a file size of 46 MB from each recorders). As a result, in this mode 2.3 gigabytes per day are accumulated (848 GBT per year).

In a fast mode (32 bands and the bandwidth of 78 kHz, time sampling 12.5 ms), $80 \times 48 \times 33 \times 4 \approx 507$ KB per second is generated from one recorder (from two recorders - 87.6 GB per day, and 32 TB per year). These data are especially valuable for the fast radio transient search, pulsars and FRBs. Fast Radio Bursts (FRB) were initially discovered in 2007 [7]. FRBs are very short transient sources of extragalactic origin, and they have a large dispersion measures (DM) >> 100 pc • cm^{-3}. To search and investigate FRBs a millisecond time scale is required.

Since June 2014, the two recorders are operating in two modes simultaneously (standard and fast). Until now (January 2017) the data of 4.5 years survey in the standard mode, and data obtained for 2.5 years of continuous observations in the fast mode have been accumulated. Totally until now about 78 TB of survey data were accumulated.

These data are the key source for Earth's ionosphere and solar flares monitoring, detection of thousands of scintillating radio sources, monitoring the flux of hundreds of radio sources in our Galaxy and beyond, wide field search for new radio pulsars and possible counterparts of cosmic gamma-ray bursts (GRB) and sources of gravitational wave transients. In the latter type of problems the most interesting is a search for FRBs. The FRBs are discovered in the S- and L-band (1–3 GHz), but in the meter wavelength (30–300 MHz) still no FRB is detected. We also believe that it possible to detect a few tens of new pulsars, and our expectations have already been justified [8, 9].

There are other aspects of these unique data, including applied research. However, in order to receive the output from this vast scientific data stream, it should be structured and efficiently processed.

2 Data Processing

2.1 BSA Database

A special database have been created in PostgreSQL in February 2014. It collects the observational data obtained from the third 128-multibeam diagram of BSA. Graphical data output is provided. The database comprises the survey data collected from July 6, 2012 up to October 20, 2013 for the standard mode [10]. Every 5 s 27 parameters were calculated for each beam, and based on these calculations, more than 700 thousand graphs were drawn, one of them is shown on Fig. 1.

In February 2014 these graphs became available on the PRAO LPI website at http://astro.prao.ru/, in December 2014 the database was launched in a public on-line access mode.

Placement of the compressed data in a relational database format allows to use all available facilities of selection, sorting, matching, filtering, and initial data reduction using standard SQL commands. This significantly simplified the cross- and temporal analysis as well as data averaging, search for data that exhibit abnormal behavior and so on.

For all of these 5 s periods (for each of 96 beams) the maximum value for the beam in the particular waveband $S_{max\_f\_n}$ (numbered as n = 1, 2,...6, where for the first band f = 109.0 MHz corresponds, for every band n = 2, ..., 6 frequency is increased on

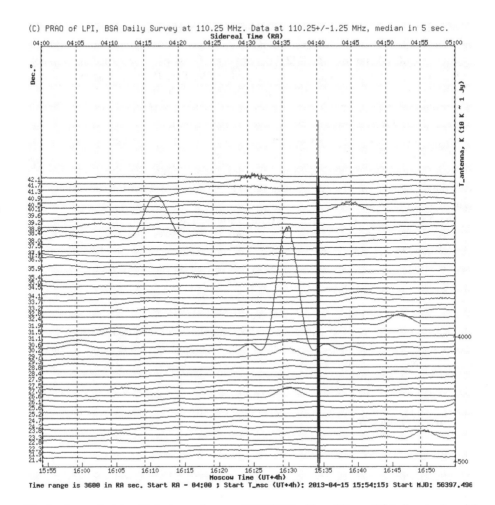

Fig. 1. A sample data of 1 h observations on 15/04/2013 with multibeam BSA diagram (special calibration signals can be seen at RA = 4 h 40 m). During passing of point-like radio sources through multibeam diagram the beam shape of the BSA diagram is clearly visible. The Y axis represents 48 beams of the BSA diagram (South - down, North - up), each beam is drawn accordingly to the Kelvin temperature scale (10 K is approximately equal to 1 Jy). X-axis (upper) is the sidereal time, and X-axis (bottom) - the time in terms of UTC + 4 h.

415 kHz, and for n = 7 it is a full waveband of 2.5 MHz), minimum $S_{min\_f\_n}$, the median average $S_{med\_f\_n}$, dispersion $N_{f\_n\ signal}$, and other parameters of 27 data types were obtained.

For calculation of the average values $S_{med\_f\_n\ Nf\_n}$ out of every 50 consecutive values 2 minimal values and 5 maximal values were discarded (to minimize influence of the short pulses presumably resulted from equipment failure); therefore, $S_{med\_f\_n\ Nf\_n}$ is calculated with remaining 43 values. Dropping of such extreme values not only makes an assessment of average values more reliable, but also allows to easily select data containing the short duration pulses (Sect. 2.2).

The database will be expanded and will go into automatic mode of filling with continuous arrival of new graphic data at the site.

2.2   Discrimination of Impulse Events

Using the database we have developed a helpful technique for the discrimination of short duration transient events, hereafter pulses, in the individual beams of the BSA diagram. The first step is logical commands in SQL, making data selection by combination of the following conditions:

- for each frequency band with number n = 1..6 we search for an overshoot at 5-second intervals with the detection criterion $S_{f\_n}/N_{f\_n} \geq 5.0$, where $S_{f\_n} = S_{max\_f\_n} - S_{med\_f\_n}$;
- condition above must be satisfied at least in 3 of the 6 frequency bands;
- testing widely separated beams of the multibeam diagram (located further than in ±3 beams). It is required that they, in contrast, show no similar overshoot, i.e. S_{f_n}/N_{f_n} must be less than 5.0. This condition rejects the majority of pulsed interferences of artificial and natural origin (e.g., thunderstorms), usually manifest themselves on all beams simultaneously. One can note that although the time accuracy of the registration of the lightning impulses does not allow the use of these data for lightning triangulation, and a joint search for the location of each lightning discharge, these data can be used for statistical analysis, for example, together with spaceborne experiments for the study of lightning activity. The pulses of astrophysical nature manifest themselves usually only in 2–3 adjacent beams of BSA diagram. Pursuing treatment this way we could misidentify extremely powerful impulse captured far side lobe and appeared more than in 3 beams.

In this manner, using the database the 83096 individual candidates in declination range $+3.5° < \delta < +42°$ during continuous observations in the period of July 2012 – October 2013 were detected [11]. Each candidate was additionally investigated using the original data file. For the two frequency bands with the maximum S/N value the Pearson correlation was performed in different time scales of 1, 2, 5, 10 s, when time series in different frequency bands shifted relative to each other. Additionally we require positive time lag, i.e. pulse in high frequency band ahead of low frequency band (which corresponds DM > 0) for secure identification of the candidate with some astrophysical source. All candidates additionally verified and sorted visually using a specially designed web interface. The results of classification are stored in a database in a special table.

All impulses with the positive lag, i.e. positive dispersion measure (DM > 0) are already sorted. An example of such an impulse (the signal from PSR2305+3100) is shown in Fig. 2.

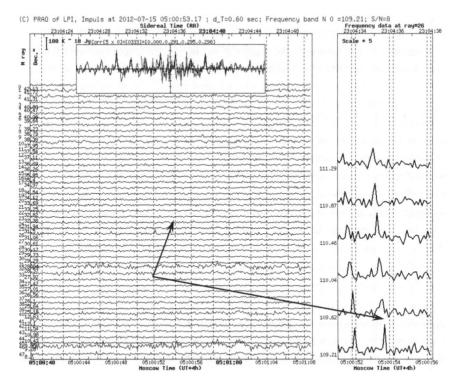

Fig. 2. An example of one of the found impulse event. The response from one of the impulse (in the center of the figure) for pulsar PSR2305+3100. Axes designations are the same as in Fig. 1. Arrows indicate the discussed impulse: (left, main box) time scan beams of the BSA multibeam diagram, (right box) - scan a beam over 6 frequencies. A shift along the frequency due to dispersion is clearly visible in the right box (DM = 49.6 pc/cm^3). The upper box shows the search result of the Pearson correlation for the four time scales: they all showed a clear shift of the lower frequency data against the high-frequency band. This pulsar is bright enough (~ 5 Jy) and the series of impulses following each other with a period of P = 1.58 s is noticeable. Totally for the PSR2305+3100 the 334 impulse events during the time interval 06.07.2012 – 20.10.2013 were detected.

A classification of candidates which do not reveal distinct time lag (scintillation, instrumental and other anthropogenic effects) has been completed. An example of such a pulse (fly-by international space station through the BSA diagram) is shown in Fig. 3.

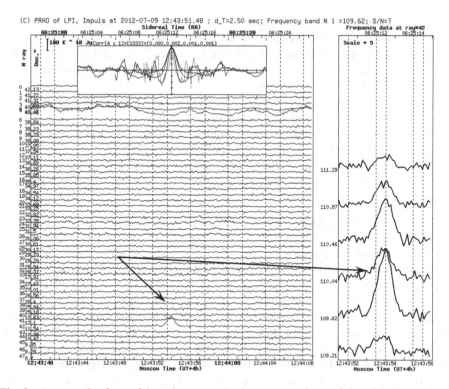

Fig. 3. An example of one of the pulse events that is a response from a flyby of the International Space Station through the radar emission beam from a Moscow region airport. These radars are operating at the frequencies around 110 MHz. Axes and other designations are the same as in the Figs. 1 and 2. The arrows indicate the selected pulse, width of the pulse about 1.5 s: in the left box a time scan of beams of the BSA diagram is presented, in the right box a time scan of the beam in 6 frequencies is shown. There is no time-frequency lag corresponding to a source that is a near Earth object. The upper figure shows the result of boxing Pearson correlation for four different time scales; the correlation coefficient is maximal for zero time lag. At least a few dozens of the events of this nature are expected in our sample. An example of the scintillation point-like radio source can be seen on the 6th beam from above in the top of the main figure. Such strong scintillation on a time scale of about 1 s is usually automatically discriminated as independent impulse events.

3 Preliminary Results

As a result a homogeneous sample of the individual impulse candidates (events) was formed to be used for further analysis of individual candidates and for statistical analysis. The preliminary results are as follow.

- (A) approximately 38% of candidates has a well-defined astrophysical nature, clearly showing the time lag corresponding to positive dispersion measure.
- (B) about 35% of candidates has the specific signature of single scintillation of radio sources of different types - from the ionosphere to the interstellar (including data from the side lobes of the diagram).
- (C) about 0.5% of the candidates are the reflection of the ground based radar emission from satellites and impulses of undefined nature.
- (D) about 13% of candidates are impulses which have anthropogenic origin: lightning, flyby airplanes through diagram beams and so on.
- (E) about 13% of candidates contain instrumental failures (some of them may still contain real impulses of undetermined origin - for example, the impulses from meteors).

The most interesting results related to this classification are due to the impulses of the astrophysical nature. In particular, from about 340 pulsars of the ATNF catalog [12], falling within a declination range $+3.5° < \delta < +42°$ (which is most suitable for our analysis) and with periods of more than 0.3 s for 41 of them we found pulses in our data. The number of individual pulses found for these pulsars ranges from one to thousands during the period of continuous observations of about one year.

For all pulsars with the dispersion measure of DM < 16, pulses are registered for 9 from the 16 pulsars (i.e. 56%). From the same ATNF pulsars with DM < 50 we registered pulses for 37 of 112 (33%), for pulsars with $50 < DM < 100$ we registered pulses for 4 from 91 (4.4%), and for DM > 100 we do not find any pulses for 187 pulsar of ATNF catalog. It is clear that pulses broadening with DM increasing is crucial for detection sensitivity of pulses from pulsars.

We are going to improve the search technique of pulse phenomena by introducing analysis of data recorded in the fast mode (time resolution of 12.5 ms) and increasing the analysis period.

We also found several groups of impulses of astrophysical origin, apparently unrelated to already known pulsars. Of these, two pulse groups are most distinguished. We assume that they are likely to be associated with two new RRAT pulsars. Preliminary parameters of the two RRAT candidates are below.

1. $R.A._{2000} = 1$ h 38 m 35 $\pm$ 30 s, $\delta_{2000} = 33° 41' \pm 15'$ and DM = 21 $\pm$ 3 pc/cm^3. For the object 34 individual impulses (in the period of July 2012 – October 2013) were found. One from the 34 impulses is shown on Fig. 4.
2. $R.A._{2000} = 10$ h 05 m 05 $\pm$ 30 s, $\delta_{2000} = 30° 10' \pm 15'$ and DM = 18 $\pm$ 2 pc/cm^3. For the object 12 individual impulses (in the period of July 2012 – October 2013) were found. One from the 12 impulses is shown on Fig. 5.

An additional check is carried out in order to exclude the possibility of the generation of these candidates by known pulsars captured in far side lobes of the BSA diagram.

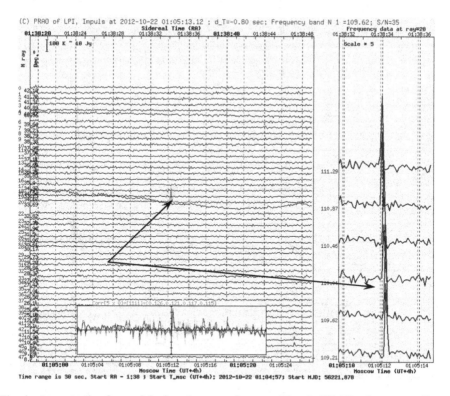

Fig. 4. An example of one of the impulse event from candidate in RRAT pulsar at coordinates R.A.$_{2000}$ = 1 h 38 m 35 ± 30 s, δ_{2000} = 33° 41′ ± 15′ and DM = 21 ± 3 pc/cm^3 for the 2012-10-12 event. Axes and other designations are the same as in the Figs. 1 and 2. The arrows indicate the selected impulse so that the left box contains a time scan of beams the of BSA diagram, the right box contains a time scan of the beam over 6 frequencies. A shift along the frequency due to dispersion is clearly visible in the right box (DM = 21 ± 3 pc/cm^3). The bottom inset shows the correlation search result. The brightness of the presented event is about 15 Jy. During continuous observations in the period of 06.07.2012 – 10.20.2013 it was detected 34 impulses from this candidate.

4   Data Access and Dissemination

A current BSA database contains observations from the BSA accumulated in a fast mode (96 beams for 6 frequency bands with a 0.1 s time resolutions) for the period of 06.07.2012 – 20.10.2013. This paper is dedicated to possibilities using the information of the database to extract transient phenomena from the BSA sky survey. All 83096 transients resulting from the extraction procedure as well as original low level data around each transient of about 30 s are also presented in the database.

Only demonstration pictures based on a pre-processed observational data are presented for a public access. View an original data and data retrieving is available only to a restricted number of investigators. In a near future a public access to the current database will be granted. We are intended for a further database development with a

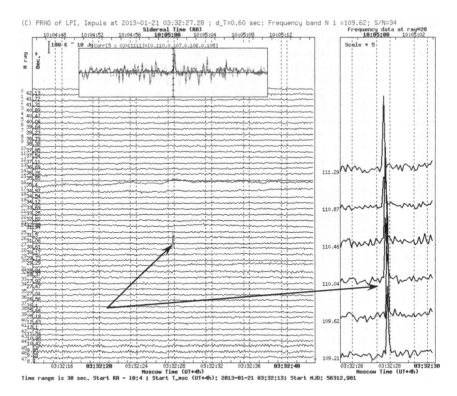

Fig. 5. An example of one of the found impulse events from possible RRAT pulsar at R. A.$_{2000}$ = 10 h 05 m 05 ± 30 s, δ_{2000} = 30° 10′ ± 15′ and DM = 18 ± 2 pc/cm^3 for the 2013-01-21 event. Axes and other designations are the same as in Figs. 1 and 2. The arrows indicate the selected impulse so that the left box contains a time scan of beams of the BSA diagram; the right box contains a time scan of the beam over 6 frequencies. A shift along the frequency due to dispersion is clearly visible in the right box (DM = 18 ± 2 pc/cm^3). The upper inset shows the Pearson correlation result for different time scales. The brightness of the presented event is about 10 Jy. During continuous observations in the period of 06.07.2012 – 10.20.2013 the 12 impulses from this candidate were detected

more data inclusion in the database and a wider access to the low level data. We are planning next steps:

– Observational data in the fast mode will be processed in a near real time and automatically stored in the database. About one hour later graphs of the processed data will be publically available at a web site of PRAO LPI.

– A data stream will be processing continuously using distributed computing based on the software platform BOINC [13]. The results will be available to all registered participants. Open access will be granted after the expiry of a proprietary time period.

– Free user database registration will be introduced. For registered users the entire database and low level data recorded in a fast mode will be available (848 GBT per year). This data is suitable for a long-term monitoring of celestial objects of various natures.

- The database will also contain all transient events found with an automatic processing. It will contain also averaged time series for different time scales and housekeeping information.
- A huge amount of low level data recorded in the fast mode (96 beams. 32 frequency bands, 12.5 ms time resolution) may not be stored completely in the on-line access at PRAO LPI servers. Instead they will be placed by parts for designated researchers, including distributed computing volunteers. All transients found in the course of processing will be stored in the database and will be available to all of the registered users after the expiry of a proprietary time period.

Acknowledgments. This work is partially supported by the RFBR grant 14-07-00870a.

References

1. Oreshko, V.V.: Radio telescopes PRAO - state and perspectives (in Russian) (2014). http://www.prao.ru/conf/rrc2014/docs/22092014/06_Oreshko.pdf
2. Van Haarlem, M.P., et al.: LOFAR: the low-frequency array. Astron. Astrophys. **556**, id. A2 (2013)
3. Konovalenko, A.A., et al.: Astronomical studies using small-size low-frequency radio telescope of new generation (in Russian). Radio Phys. Radio Astron. **21**(2), 83–131 (2016)
4. Braude, S.Y., Men, A.V., Sodin, L.G.: UTR-2 radio telescope for the decameter wave range. Antenny, no. 26, pp. 3–15 (1978)
5. Taylor, G.B., et al.: First light for the first station of the long wavelength array. J. Astron. Instrum. **1**, 1–56 (2012)
6. Coenen, T., et al.: The LOFAR pilot surveys for pulsars and fast radio transients. Astron. Astrophys. **570**, id. A60, 16 pp. (2014)
7. Lorimer, D.R., et al.: A bright millisecond radio burst of extragalactic origin. Science **318** (5851), 777–780 (2007)
8. Tyul'bashev, S.A., Tyul'bashev, V.S.: The discovery of new pulsars on the BSA LPI radio telescope. Astron. Tsirkulyar **1624**, 1–4 (2015)
9. Rodin, A.E., Samodourov, V.A., Oreshko, V.V.: Fast method of detection of periodical radio sources. Astron. Tsirkulyar **1629**, 1–4 (2015)
10. Samodurov, V.A., Rodin, A.E., Kitaeva, M.A., Isaev, E.A., Dumsky, D.V., Churakov, D.D., Manzyuk, M.O.: The daily 110 MHz sky survey (BSA FIAN): online database, science goals data processing by distributed computing. In: Proceedings of the Data Analytics and Management in Data Intensive Domains (DAMDID-2015), pp. 127–128 (2015)
11. Samodurov, V.A., Pozanenko, A.S., Rodin, A.E. Churakov, D.D., Dumsky, D.V., Isaev, E. A., Kazantsev, A.N., Logvinenko, S.V., Oreshko, V.V., Toropov, M.O., Volobueva, M.I.: The daily radio sky survey at 110 MHz: database and statistical analysis of transient phenomena in 2012–2013. In: Selected Papers of the XVIII International Conference on Data Analytics and Management in Data Intensive Domains, vol. 1752, pp. 213–218. CEUR Workshop Proceedings (2016)
12. Hobbs, G., Manchester, R.N., Toomey, L.: ATNF Pulsar Catalogue v1.54. http://www.atnf.csiro.au/people/pulsar/psrcat/
13. BOINC: Open-source software for volunteer computing. http://boinc.berkeley.edu/

Data Analysis

Semantics and Verification of Entity Resolution and Data Fusion Operations via Transformation into a Formal Notation

Sergey Stupnikov[✉]

Institute of Informatics Problems, Federal Research Center
"Computer Science and Control" of the Russian Academy of Sciences,
Moscow, Russia
sstupnikov@ipiran.ru

Abstract. During all the period of development of data integration methods and tools the issues of formal semantics definition and verification were arising. Three levels of integration can be distinguished: data model integration, schema matching and integration and data integration proper. This paper is aimed at development of methods and tools for formal semantics definition and verification on the third level – level of data proper. An approach for definition of formal semantics for high-level data integration programs is proposed. The semantics is defined using a transformation into a formal specification language supported by automatic/interactive provers. The semantics is applied for verification of structured data integration workflows. Workflow properties to be verified are presented as expressions of the specification language chosen. After that a semantic specification of the data integration workflow is verified w.r.t. required properties. A practical aim of the work is to define a basis for formal verification of data integration workflows during problem solving in various integration environments.

Keywords: Entity resolution · Data fusion · Formal semantics and verification · Model transformation

1 Introduction

Due to significant growth of volume and diversity of data nowadays the development of methods and tools for data integration is becoming more and more urgent. Data with diverse structure and semantics can be required for solving the same scientific or industrial problem. Increasing data volume requires scalable platforms for distributed parallel data processing like Apache Hadoop [4]. Obviously all components of data integration process should be implemented as programs over the respective platform.

The following levels of integration (from higher to lower) can be distinguished: data model integration [18, 19], schema matching and integration [30] and data integration proper. Usually completion of the integration on a higher level is a prerequisite for integration on a lower level. Note that this work considers integration of *structured* data. Such data conform some schema defining types (structures) of data.

© Springer International Publishing AG 2017
L. Kalinichenko et al. (Eds.): DAMDID/RCDL 2016, CCIS 706, pp. 145–162, 2017.
DOI: 10.1007/978-3-319-57135-5_11

On the level of data models elements of models (languages) are matched. On the level of schemas data types (structures) and their attributes are matched. On the level of data proper rules for transformation of data collections, their elements and attribute values are defined. As an ideal definition of data integration process one can consider a set of declarative rules (like a Datalog program). Such approach is called *data exchange* [11]. In practice data integration process is a set of data transformation operations represented in SQL or SQL-like language, and order of operations execution is essential. Usually data integration is organized as a workflow. In both data exchange and workflow cases data integration is conformed with the previous level of integration, i.e. schema matching.

Data integration processes are quite important and complicated. Obviously during the period of development of data integration methods and tools the issues of formal semantics definition and verification were arising.

Methods and tools for data model unification preserving semantics had been developed [17–20, 34, 35] (data model level) as well as methods and tools for compositional development of information systems and its verification [31, 33] (schema level). Hadoop-based advanced environments for data integration and problem solving over heterogeneous resources are being developed [21, 22, 36]. A short review of mentioned and related work is provided in Sect. 2.

This paper is aimed at development of methods and tools for formal semantics definition and verification on the third level – level of data proper. It is assumed that data model integration and schema matching have been already completed. It is also assumed w.r.t. practical reasons that data integration process is defined as a set of operations expressed in SQL-like language and not as Datalog-like program. Properties of integration process as a whole are not defined explicitly, only rules for data transformation in concrete operations are defined.

Several groups of operations constituting a data integration process can be distinguished: data transformation, entity resolution [13, 23] and data fusion [6, 7, 10]. *Data transformation* implements movement from a source schema into the target (unified, integrated) schema. *Entity resolution* is an extraction and binding of information describing the same real word entity from different data collections [23]. *Data fusion* is a combination of different representation of the same real world entity into unified representation [28].

So, a data integration process is a workflow; the activities of the workflow are separate operations of transformation of structured data, entity resolution and data fusion. Several generic data fusion operations are distinguished in [7].

To apply data transformation, entity resolution and data fusion methods over distributed data processing platforms specific high-level programming languages are being developed. Examples of these languages are Jaql [5], Pig Latin [29], Highlevel Integration Language (HIL) [15]. Distributed execution of programs is provided via compilation of high-level programs into imperative language (like Java) programs and further execution using MapReduce computation model [27].

The idea of the approach proposed in this paper is the following. A high-level language intended for data transformation, entity resolution and data fusion is mapped into a formal specification language supported with tools for automatic and/or interactive proof.

In this way high-level data integration programs (workflows) gain formal semantics. Properties of programs to be verified are represented as expressions of the formal language. After that a specification providing semantics for a data integration workflow is verified to conform these properties using formal proof facilities.

To illustrate the approach the HIL is used as a high-level data integration language. The language is developed by IBM and supplied as a part of Hadoop-based solution BigInsights 3.0 [14] and as a part of InfoSphere Big Match for Hadoop [37]. HIL was successfully applied in projects of financial and social data integration [15]. Scalability of HIL w.r.t. data volume as well as schema diversity was demonstrated. Comparing with traditional tools for ETL-processes construction based on relational data model and SQL, HIL provides much more flexible declarative data integration facilities aimed at entity resolution and data fusion.

As a formal specification language the Abstract Machine Notation (AMN) [1] is used. AMN is based on set theory and typed first order predicate logic. AMN specifications are called *abstract machines* combining state space (a set of variables) and behavior definitions. Behavior is defined by operations over state variables. The language is supported with tools for formal proof [2]. The choice of AMN is motivated by capabilities of the language and experience of the author in the area of usage AMN for definition of formal semantics. To formalize semantics of HIL process specification languages like finite automata or Petri nets are not sufficient. It is required also to specify complicated data transformation operations which are activities constituting workflows. Some languages providing similar capabilities like RAISE or Z can be applied instead of the AMN.

The structure of the paper is as follows. Section 2 considers related work in the field of definition of formal semantics and verification of data integration process on different levels. Section 3 considers semantic mapping of HIL into AMN and some issues of implementation of the mapping. Section 4 illustrates verification of integration workflow properties by an example.

This work is a continuation of [39]. Comparing it with the previous work in [39], the main section of the paper (Sect. 3) is significantly extended; mapping of HIL into AMN is formalized using semantic functions. Examples are extended. The mapping is also implemented using ATLAS Transformation Language (ATL) [3], some implementation issues are considered. A link to the repository with the source code and samples [38] is provided.

2   Related Work

Works devoted to definition of semantics of conceptual modeling languages and data models are known since late 1990s. In [16] a mapping of ODMG'93 relationship type into AMN was used for verification of mapping the relationship type into the canonical data model. The intension was to verify preservation of semantics by data model mapping.

Some works of Lano, Bicarregui (e.g. [25]) are devoted to formal definition of UML in RAL (Real-time Action Logic). The ways of applying semantics for verification of transformations of UML diagrams are provided. Examples of transformations are strengthening the invariants, elimination of many-to-many associations, aggregation transitivity, interface transformation and others.

In [33] semantics of an object data model in AMN is defined for verification of compositional development of information systems. Semantics can be used for formal proof that a composition of legacy program components can be used as an implementation of an abstract specification of an information system to be developed.

Later AMN was used for verification of mappings of various kinds of data models: process models [17], ontological models [20], array models [34], graph models [35].

Methods and tools for data model unification (mapping of a data model into the unified canonical data model) were developed [18, 19]. Verification of data model mappings was also based on representation of data model semantics in AMN.

A lot of works are known in the field of *model transformation* verification [9, 26]. These works apply the Model-Driven Engineering (MDE) methodology. Within MDE "a model" means both a data model (language) and a conceptual schema. Model transformation is an implementation of model mapping. A transformation is a set of rules defining the way to transform entities of a source model into entities of a target model.

Properties of model transformations to be verified are *termination* (the transformation execution finishes for any well-formed transformation specification), *determinism* (uniqueness of the target model for a given source model and transformation specification), *syntactic correctness* (generated model is a syntactically well-formed instance of the target language), *preservation of execution semantics* (transformation execution must behave as expected according to the definition of the transformation language semantics). To provide a possibility of formal proof the properties of transformations are represented in different variants of first order logic languages (like Calculus of Inductive Constructions). Depending on the power of formal language automatic provers (SAT solvers, model checkers) of automated logic inference can be applied.

Several approaches for definition of formal semantics of data transformation are known. For instance, in [24] formal semantics of data flow diagrams in VDM (Vienna Development Method) is considered. In [32] semantics of ETL-processes is represented in LDL (Logic-Based Data Language). The aim of these works is to provide a transition from a conceptual model of data transformation process to a logical model. Clio system [12] implements *data exchange* approach assuming data transformation as a Datalog-like program possessing logical semantics.

The main features of this particular paper are as follows. Formal semantics in AMN is provided for high-level entity resolution and integration language (HIL). Semantic mapping is implemented as a transformation in ATL [3]. Semantics defined is applied to verification of data integration workflows. Main features of the approach are considered in Sects. 3 and 4.

3 Formal Semantics of the Entity Resolution and Integration Language

The approach for the definition of formal semantics of the HIL in AMN is illustrated in this section by an example of data integration workflow. This example of financial data integration was originally considered in works devoted to Midas integration system [8] and HIL [15]. Financial data used is published by U.S. Securities and Exchange Commission.

The example considered is just a part of the whole financial data integration workflow. The aim of the example workflow is to create the collection of entities (records) named *Person*. The entities of the collection correspond to managers of leading US companies. Source data collections for the workflow are *InsiderReportPerson (IRP)* и *JobChange*. The first source collection contains data extracted from insider salary reports. The second source collection contains data extracted from hiring and job change reports. The extraction of the information from unstructured data is not an issue of this paper. The workflow includes five operations:

- *insertIntoPositionsFromIRP* (extract data concerning positions of managers from the *IRP* collection and insert respective records into auxiliary collection *Positions*);
- *createPeopleLink* (create *PeopleLink* collection of relationships between managers and job change records);
- *insertIntoPositionsFromJobChangePeopleLink* (extract data concerning positions of managers from the *JobChange* collection and insert respective records into *Positions* collection);
- *insertIntoEmploymentFromJobChangePeopleLink* (insert hiring and job change data into auxiliary collection *Employment*);
- *insertIntoPersonFromIRP* (insert data concerning managers and their positions into *Person* collection).

Among these operations *createPeopleLink* is a typical example of entity resolution, both *insertIntoEmployment* and *insertIntoPositionsFromJobChange* operations are examples of data fusion.

Note that separate operations (called *statements* or *rules* [15]) according to HIL syntax [37] do not possess names like in SQL. Names mentioned are assigned to operations for convenience of reasoning.

Note also that HIL programs do not presume intrinsic order among operations. "It is the role of the compiler to stage the execution so that any intermediate entities are fully materialized before they are used in other entities" [15]. For instance, operation *insertIntoEmployment* should be executed after *insertIntoPositionsFromIRP, createPeopleLink, insertIntoPositionsFromJobChange* operations since it uses *PeopleLink* and *Positions* collections, but these collections are populated by next three operations mentioned.

Semantics of basic constructions of the HIL in AMN is defined below and illustrated by examples. Semantic mapping is defined using a set of semantic functions. According to principles of denotational semantics, semantic functions map elements of *syntactic domains* of HIL into constructions of AMN. Syntactic domains of HIL are

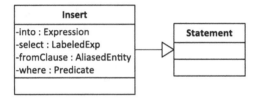

Fig. 1. Abstract syntax domains of HIL

structured applying *Ecore* metamodel (an implementation of OMG's Essential Meta-Object Facility) used in Eclipse Modeling Framework [40]. *HIL.ecore* model [38] conforming *Ecore* metamodel was developed. In fact, this model represents *abstract syntax* of HIL neglecting any syntactic sugar. For instance, insert statement is formalized as *Insert* syntax domain (*Ecore* class) (Fig. 1). *Insert* is a subdomain of the *Statement* domain. Elements of *Insert* domain (insert statements) include *into, select from, where* clauses formalized as *Expression, LabeledExp, AliasedEntity, Predicate* syntax domains respectively.

Concrete syntax of HIL [38] binding syntactic sugar and abstract syntax was formalized using EMFText framework [41]. For instance, *Insert* syntactic domain is bound with syntactic sugar in the following way:

```
Insert ::=
    "insert" "into" into
    "select" "[" select ("," select)* "]"
    "from" fromClause ("," fromClause)*
    ("where" where)? ";" ;
```

Nonterminal symbols of the concrete grammar correspond with syntactic domains, symbol "*" means repetition, symbol "?" means optionality.

In this section semantic function names start with "s" symbol, parameters of semantic functions are embraced by square brackets. Semantic functions, syntax domains, metavariables are styled with *italic*. Terminal HIL and AMN grammar symbols are styled with roman.

3.1 Semantics of HIL Programs

Every HIL program is represented by a separate REFINEMENT [1] construction. These kind of abstract machines provide the most comprehensive specification capabilities. Semantic function for the *Program* domain is defined in the following way:

```
sProgram[Statement*] ≙
REFINEMENT HILProgramSemantics
ABSTRACT_CONSTANTS sAbstractConstants[Statement*]
PROPERTIES sProperties[Statement*]
VARIABLES sVariables[Statement*]
INVARIANT sInvariant[Statement*]
INITIALISATION sInitialisation[Statement*]
OPERATIONS sOperations[Statement*]
END
```

This function forms semantic specification for a whole program. Particular sections of specification (like variables or operations sections) are formed using particular semantic functions defined below.

3.2   Semantics of Entity Type, Index Type and Function Declarations

Entity and index types of HIL are represented in AMN as variables in VARIABLES section; variables are typed in INVARIANT section and initialized in INITIALISATION section.

For instance, *entity type IRP* is represented in AMN in the following way:

HIL construction	AMN construction

declare IRP: set[
  name: string,
  cik: int,
  bdate: string;
  company: string,
  title: string,
  isOfficer: Boolean,
  isDirector: Boolean];
	VARIABLES IRP, …
	INVARIANT
	  IRP: POW(struct(
	    name: STRING_TYPE, cik: INT,
	    bdate: STRING_TYPE, company: STRING_TYPE,
	    title: STRING_TYPE, isOfficer: BOOL,
	    isDirector: BOOL
	  )) & …
	INITIALISATION
	  IRP := {} || …

For the *IRP* type a variable with the same name is declared. The variable is typed as a subset (*POW*) of the set of tuples (*struct*). Tuples include attributes corresponding to the attributes of the *IRP* type. One-to-one correspondence is established among built-in types of AMN and HIL. For instance, *string* type of HIL corresponds to *STRING_TYPE* of AMN. These mappings are formalized with the following semantic functions:

$$sVariables[\text{ declare } name: type; \text{ }] \triangleq name$$
$$sInitialisation[\text{ declare } name: type; \text{ }] \triangleq name := \{\}$$
$$sInvariant[\text{ declare } name: type; \text{ }] \triangleq name: sType[type]$$
$$sType[\text{ set } type \text{ }] \triangleq POW(sType[type])$$
$$sType[\text{ ``[" } element* \text{ ``]" }] \triangleq sRecordElement[element*]$$
$$sRecordElement[name: type] \triangleq name: sType[type]$$
$$sType[string] \triangleq STRING\_TYPE$$
$$sType[int] \triangleq INT$$
$$sType[Boolean] \triangleq BOOL$$

Here semantic functions take generalized syntactic definitions of HIL as parameters. For instance, *sVariables* function takes syntactic expression of *Declaration* syntax domain. A declaration syntactically looks like "declare *name: type*;". Words of roman type (like "declare") are HIL keywords. Words of italic type (like "*name*" and "*type*") are metavariables having values of other syntax domains (for instance, *type* is of *Type* syntax domain).

Index type (fmap) in HIL is a mapping of instances of index *key* type into instances of index *value* type. For instance, index type *Positions* is represented in AMN in the following way:

HIL construction	AMN construction

declare Positions:
fmap [cik: int,
  company: string]
to set [title: string];
	VARIABLES Positions, …
	INVARIANT
	  Positions: struct(cik: INT,
	    company: STRING_TYPE) +->
	    POW(struct(title: STRING_TYPE))
	INITIALISATION
	  Positions := {} || …

To map *fmap* HIL construction into AMN the partial function constructor "+->" is used. Semantic function *sType* is defined on *fmap* construction in the following way:

$$sType[\texttt{fmap}\ fromType\ \texttt{to}\ toType] \triangleq sType[fromType]\ \texttt{+->}\ sType[toType]$$

HIL *function* declarations are represented in AMN as *abstract constants* with the same name. Constants are typed in PROPERTIES section of an abstract machine. For instance, *compareName* HIL function is represented in AMN in the following way:

HIL construction	AMN construction

declare compareName:
  function
    string, string
    to boolean;
	ABSTRACT_CONSTANTS compareName, …
	PROPERTIES
	  compareName:
	  ((STRING_TYPE * STRING_TYPE) --> BOOL)

This mapping is formalized using semantic functions as follows:

$$sAbstractConstants[\texttt{declare}\ name:\ \texttt{function}\ fromType*\ \texttt{to}\ toType] \triangleq name$$
$$sProperties[\texttt{declare}\ name:\ \texttt{function}\ fromType1,\ fromType2\ \texttt{to}\ toType] \triangleq$$
$$name:\ ((sType[fromType1]*sType[fromType2])\ \texttt{-->}\ sType[toType])$$

3.3 Semantics of HIL Statements

Every HIL statement is represented in AMN as an operation of abstract machine.

3.3.1 Semantics of Entity Resolution Statements

As an example of such statement *createPeopleLink* is considered. The statement is represented in AMN in the following way:

HIL construction	AMN construction
| ```
create link PeopleLink as
select [ cik: pp.cik,
   docID: jj.docID,
   span: jj.span ]
from IRP pp, JobChange jj,
   pp.emp ee
match using
rule1:
   (ee.company = jj.company) and
   (compareName(pp.name,jj.name)=
       true),
rule2: (normName(pp.name) =
          normName(jj.name))
check
   if not(null(jj.bdate))
       and not(null(pp.bdate))
   then (jj.bdate = pp.bdate);
``` | ```
createPeopleLink =
SELECT  TRUE = TRUE
THEN PeopleLink :=
{rr | #(pp, jj, ee).(pp : IRP &
   jj : JobChange & ee : pp'emp &
   rr = rec(cik: pp'cik,
      docID: jj'docID, span: jj'span) &
   (ee'company = jj'company &
   compareName(pp'name, jj'name) =
     TRUE or
   normName(pp'name)=normName(jj'name))&
   (not((jj'bdate = null_string)) &
   not((pp'bdate = null_string)) =>
       jj'bdate = pp'bdate)))};
   state'PeopleLinkCreated := TRUE
END
``` |

The structure of entities of collection *PeopleLink* to be created is defined in *select* clause of the statement. Data are extracted from *IRP* and *JobChange* collections and joined (*from* clause). *Match using* clause specifies the possible ways in which the input entities can be matched. In the example two ways labeled as *rule1* and *rule2* are provided. The set of links that are produced by the *match using* clause are called *candidate links* [15]. The *check* clause specifies further predicates that must be satisfied by all output links.

The basic ideas of the mapping are as follows. *PeopleLink* variable is assigned with a set created by the set comprehension construction $\{rr \mid F(rr)\}$. Here rr is the set element variable and $F(rr)$ is the comprehension predicate. Variable rr is typed with the record type *rec*. Attributes of the record are the attributes of *PeopleLink* collection type. Variable (*pp, jj, ee*) is created for every collection to be joined. Variables are bound with the existential quantifier and typed appropriately: *pp: IRP & jj: JobChange & ee: pp'emp*. Conditions from *match* and *check* clauses are represented as predicates over *pp, jj, ee* variables.

The mapping is formalized using semantic functions in the following way:

```
sOperations[ create link name as
  select select* from fromClause*
  match using match1, match2
  check check1 check2 ] ≙
sCreateLinkOperationName[name] =
SELECT sStateCheckPredicate[name]
THEN
  name := { rr | #(sExistentialVariables[fromClause*]).(
    sExistVarTyping[fromClause*] &
    rr = rec(sRecord[select*]) &
    (sMatch[match1] or sMatch[match2]) &
    sCheck[check1] & sCheck[check2]
  )};
  sStateSetSubstitution[name]
END
```

```
sCreateLinkOperationName[name] ≙ concatenateStrings("create", name)
sExistentialVariables[exp alias] ≙ alias
sExistVarTyping[exp alias] ≙ alias: sExpression[exp]
sRecord[label: exp] ≙ label: sExpression[exp]
sMatch[name: predicate] ≙ sPredicate[predicate]
sCheck[if ifPred then thenPred] ≙
    sPredicate[ifPred] => sPredicate[thenPred]
sPredicate[predicate1 and predicate2] ≙
    sPredicate[predicate1] & sPredicate[predicate2]
sPredicate[not(null(exp))] ≙ not(sExpression[exp] = null_string)
sPredicate[exp1 = exp2] ≙ sExpression[exp1] = sExpression[exp2]
sExpression[Identifier] ≙ Identifier
sExpression[recordID.elementId] ≙ recordID'elementId
sExpression[functionId(parameter1, parameter2)] ≙
    functionId(sExpression[parameter1], sExpression[parameter2])
```

Semantic functions for state manipulating (*sStateCheckPredicate, sStateSetSubstitution*) are defined below in Subsect. 3.4. Auxiliary function *concatenateStrings* is just string concatenation.

3.3.2   Semantics of Entity Population Statements

As an example of such statement *insertIntoEmploymentFromJobChangePeopleLink* is considered. The statement is represented in AMN in the following way:

HIL construction	AMN construction
| ```
insert into
  Employment!
    [cik: LL.cik]
select
[ company:
    jj.company,
  positions:
    Positions!
    [cik: LL.cik,
     company:
       jj.company]
]
from
  JobChange jj,
  PeopleLink LL
where
(jj.docID= LL.docID)
  and
(jj.span = LL.span);
``` | ```
insertIntoEmploymentFromJobChangePeopleLink =
SELECT state'PeopleLinkCreated = TRUE &
  state'
  PositionsFromJobChangePeopleLinkInserted =
    TRUE &
  state'PositionsFromIRPInserted = TRUE
THEN
  Employment := (Employment \/
  {rr1 | #(jj1, LL1).(jj1 : JobChange &
  LL1 : PeopleLink &
  rr1 = (rec(cik: LL1'cik) |->
    {rr | #(jj, LL).(jj : JobChange &
      LL : PeopleLink &
      rr = rec(company: jj'company,
        positions: Positions(
          rec(cik: LL'cik,
            company: jj'company))) &
      LL'cik = LL1'cik & jj'docID = LL'docID &
      jj'span = LL'span)})))});
  state'
  EmploymentFromJobChangePeopleLinkInserted :=
  TRUE
END;
``` |

Sign "!" after the identifier of the populated collection *Employment* of index type means that *cik* attribute is considered as key of index type elements. Collection populating is represented in AMN by set union operation "\/". *Employment* collection is populated by a set of entities created by the set comprehension construction (as in previous Subsect. 3.3.1) using *rr1* variable. The variable is typed with the pair type. The first element of the pair is the record *rec(cik: LL1'cik)* corresponding to the key of the *Employment* index type. The second element of pair is a set constructed in the same way as set in Subsect. 3.3.1. Elements of the pair are combined with "|->" symbol.

The mapping is formalized using semantic functions in the following way:

```
sOperations[ insert into name!exp as
  select select* from fromClause*
  where predicate ] ≙
concatenateStrings("insertInto", name, "From",
  sInsertOperationName[fromClause*]) =
SELECT sStateCheckPredicate[name fromClause*]
THEN
  name := name \/ { rr1 | #(sExistentialVariables1[fromClause*]).(
    sExistVarTyping1[fromClause*] &
    rr1 = rec(sRecord1[exp]) |->
    { rr | #(sExistentialVariables[fromClause*]).(
            sExistVarTyping[fromClause*] & rr = rec(sRecord[select*]) &
            sEqualityPredicate[exp] & sPredicate[predicate]) }
  )};
  sStateSetSubstitution[name fromClause*]
END

sInsertOperationName[Identifier1 alias1, Identifier2 alias2] ≙
  concatenateStrings(Identifier1, Identifier2)
sExistentialVariables1[exp alias] ≙ concatenateStrings(alias, "1")
sExistVarTyping1[exp alias] ≙
  concatenateStrings(alias, "1"): sExpression[exp]
sRecord1[label: exp] ≙ label: sExpression1[exp]
sExpression1[recordID.elementId] ≙
  concatenateStrings(recordID, "1")'elementId
sEqualityPredicate[exp] ≙ sExpression[exp] = sExpression1[exp]
```

3.4   Operation Execution Order Semantics

For tracking execution of operations in *HILProgramSemantics* machine the *state* variable is added to VARIABLES section:

```
VARIABLES … state, …
INVARIANT
state: struct(
  PeopleLinkCreated: BOOL,
  PositionsFromIRPInserted: BOOL,
  PositionsFromJobChangePeopleLinkInserted: BOOL,
  EmploymentFromJobChangePeopleLinkInserted: BOOL,
  PersonFromIRPInserted: BOOL
) & …
INITIALISATION
  state:= rec(PeopleLinkCreated: FALSE,
    PositionsFromIRPInserted: FALSE,
    PositionsFromJobChangePeopleLinkInserted: FALSE,
    EmploymentFromJobChangePeopleLinkInserted: FALSE,
    PersonFromIRPInserted: FALSE) || …
```

The variable is typed in INVARIANT with *struct* type. Attributes of the type correspond to operations of the machine. Every attribute is initialized with *FALSE* value. When an operation is executed, the value of the respective attribute is set to *TRUE*. For instance, *state'PeopleLinkCreated* is set to *TRUE* when *CreatePeopleLink* is executed (Subsect. 3.3.1).

Semantic functions that create typing predicate, initialization for *state* variable and state assignment substitution are defined in the following way:

sStateTypingPredicate[*Statement1 Statement2*] ≜
  state: struct(*sStateAttributeName*[*Statement1*]: BOOL,
                sStateAttributeName[*Statement2*]: BOOL)
sStateAttributeName[create link *name* as
  select *select** from *fromClause** match using *match** check *check**] ≜
  concatenateStrings(*name*, "Created")
sStateAttributeName[insert into *name*!*exp* as
  select *select** from *fromClause** where *predicate*] ≜
  concatenateStrings(*name*, "From", *sInsertOperationName*[*fromClause**],
    "Inserted")
sStateInitialisation[*Statement1 Statement2*] ≜
  state:= rec(*sStateAttributeName*[*Statement1*]: FALSE,
              sStateAttributeName[*Statement2*]: FALSE)
sStateSetSubstitution[*name fromClause**] ≜
  state'*sStateAttributeName*[*name fromClause**] := TRUE

Every operation is executed only if its precondition (predicate in SELECT clause) is evaluated to *TRUE*. For instance, precondition for the *insertIntoEmployment FromJobChangePeopleLink* operation is execution of *insertIntoPositionsFromIRP*, *createPeopleLink, insertIntoPositionsFromJobChange* operations. Precondition predicate looks as follows (Subsect. 3.3.2):

```
state'PeopleLinkCreated = TRUE &
state'PositionsFromJobChangePeopleLinkInserted = TRUE &
state'PositionsFromIRPInserted = TRUE
```

Precondition predicates are constructed using *precedingStatements* function. The function maps every HIL statement to a set of preceding statements. The function is calculated in the following way: for any statement s, a statement t belongs to the set *precedingStatements(s)* if and only if a collection c exists such that c is created or populated by t and c is used in *from* or *select* clauses of s. Collection c must be materialized before s is executed; and c is populated in t; so t must be executed before s.

Semantic function that create precondition predicate is defined in the following way:

```
sStateCheckPredicate[Statement] ≜
    precedingStatements(Statement) = {s₁, … , sₙ} ⇒
    state'sStateAttributeName[s₁] = TRUE & … &
    state'sStateAttributeName[sₙ] = TRUE
```

Here $P \Rightarrow E$ is an auxiliary construction of denotational semantics having a value of expression E when predicate P holds.

Semantic functions defined in this section were implemented [38] using ATLAS Transformation Language (ATL) [3]; a declarative/imperative transformation of HIL programs into AMN specifications was created. The most part of the mapping was implemented using declarative rules. Some features of mapping (like *precedingStatements* function) were implemented using imperative statements. Note that transformation supports a subset of HIL language. Several features of *create link* statements are not covered: *block, score, group on, cardinality* [37].

4   Application of Formal Semantics for Verification of Data Integration Workflows

According to the proposed approach for a HIL program of data integration workflow to be verified a semantic specification is generated automatically using ATL transformation [38]. Then properties of workflow to be verified are represented as AMN predicates. These predicates are added to INVARIANT section of the semantic specification.

After that semantic specification is loaded into Atelier B [2] tool. Syntax and type checking of the specification are done automatically. The aim is to prove that invariant is preserved by operations. Atelier B automatically generates all the required theorems. The theorems are to be proved using automatic and interactive provers.

As an example consider a property of the workflow presented in the previous section. The property is represented as an AMN predicate:

```
(state'PersonFromIRPInserted = TRUE =>
  !(nn, cc, tt).(
    (#(irp).(irp: IRP & nn = irp'name &
        cc = irp'company & tt = irp'title) or
    #(jc).(jc: JobChange & nn = jc'name &
        cc = jc'company & tt = jc'appointedAs)) =>
    #(pers).(pers: Person & pers'name = normName(nn) &
    #(ee).(ee: pers'emp & ee'company = cc &
      #(pos).(pos: ee'positions & pos'title = normTitle(tt)) ))))
```

The property states that upon completion of the workflow (the final operation is *insertIntoPersonFromIRP*) the following holds. For every combination of a person name, a company name, and a position title that can be found in *IRP* or *JobChange* collections there exists a record with the same person name, company name, and position title (up to normalization) in *Person* collection. This means that the workflow extracts all information concerning positions of persons from source collections.

HIL program of the example considered in Sect. 3 was automatically transformed into AMN specification using transformation [38]. The property was added into INVARIANT section, final specification was loaded into Atelier B. Syntax and typing of the specification were checked. Invariant preserving theorems were generated.

5   Conclusions

The paper considers formal semantics of the HIL language intended for description of complicated data integration workflows including data transformation, entity resolution and data fusion operations. Semantics of HIL programs is represented in AMN formal specification language. Mapping of the HIL into AMN is formalized using a set of semantic functions. The mapping is implemented as a transformation in ATLAS Transformation Language. The implementation opens a way to use formal verification of data integration workflow properties during problem solving in different integration environments [21, 22, 36].

Acknowledgments.   This research was partially supported by the Russian Foundation for Basic Research (projects 15-29-06045, 16-07-01028).

References

1. Abrial, J.-R.: The B-Book: Assigning Programs to Meanings. Cambridge University Press, Cambridge (1996)
2. Atelier B, the industrial tool to efficiently deploy the B Method. http://www.atelierb.eu/
3. ATL - a model transformation technology. https://eclipse.org/atl/
4. Apache Hadoop Project (2016). http://hadoop.apache.org/

5. Beyer, K.S., Ercegovac, V., Gemulla, R., Balmin, A., Eltabakh, M., Kanne, C.-C., Ozcan, F., Shekita, E.J.: Jaql: a scripting language for large scale semistructured data analysis. In: 37th International Conference on Very Large Data Bases VLDB, pp. 1272–1283. Curran Associates, New York (2011)

6. Bleiholder, J., Naumann, F.: Data fusion. ACM Comput. Surv. **41**(1). Article No. 1. (2009). doi:10.1145/1456650.1456651

7. Bleiholde, J.: Data fusion and conflict resolution in integrated information systems. D.Sc. Diss., 184 p., Hasso-Plattner-Institut, Potsdam (2010)

8. Burdick, D., Hernández, M.A., Ho, H., Koutrika, G., Krishnamurthy, R., Popa, L., Stanoi, I. R., Vaithyanathan, S., Das, S.: Extracting, linking and integrating data from public sources: a financial case study. IEEE Data Eng. Bull. **34**(3), 60–67 (2011)

9. Calegari, D., Szasz, N.: Verification of model transformations: a survey of the state-of-the-art. Electronic Notes in Theoretical Computer Science **292**, 5–25 (2013)

10. Luna Dong, X., Naumann, F.: Data fusion — resolving data conflicts in integration. Proc. VLDB Endowment **2**(2), 1654–1655 (2009)

11. Fagin, R., Kolaitis, P., Miller, R., Popa, L.: Data exchange: semantics and query answering. Theoret. Comput. Sci. **336**(1), 89–124 (2005)

12. Fagin, R., Haas, L.M., Hernández, M., Miller, R.J., Popa, L., Velegrakis, Y.: Clio: schema mapping creation and data exchange. In: Borgida, A.T., Chaudhri, V.K., Giorgini, P., Yu, E.S. (eds.) Conceptual Modeling: Foundations and Applications. LNCS, vol. 5600, pp. 198–236. Springer, Heidelberg (2009). doi:10.1007/978-3-642-02463-4_12

13. Getoor, L., Machanavajjhala, A.: Entity resolution for big data. In: KDD 2013: 19th ACM SIGKDD Conference on Knowledge Discovery and Data Mining Proceedings, pp. 1527–1527 (2013)

14. IBM InfoSphere BigInsights Version 3.0 Information Center. https://goo.gl/lZpEQd

15. Hernandez, M., Koutrika, G., Krishnamurthy, R., Popa, L., Wisnesky, R.: HIL: a high-level scripting language for entity integration. In: 16th Conference (International) on Extending Database Technology Proceedings EDBT 2013, pp. 549–560 (2013)

16. Kalinichenko, L.A.: Method for data models integration in the common paradigm. In: Proceedings of the First East-European Symposium on Advances in Databases and Information Systems ADBIS 1997, vol. 1: Regular Papers, pp. 275–284. Nevsky Dialect, St.-Petersburg (1997)

17. Kalinichenko, L., Stupnikov, S., Zemtsov, N.: Extensible canonical process model synthesis applying formal interpretation. In: Eder, J., Haav, H.-M., Kalja, A., Penjam, J. (eds.) ADBIS 2005. LNCS, vol. 3631, pp. 183–198. Springer, Heidelberg (2005). doi:10.1007/11547686_14

18. Kalinichenko, L.A., Stupnikov, S.A.: Constructing of mappings of heterogeneous information models into the canonical models of integrated information systems. In: Advances in Databases and Information Systems: Proceedings of the 12th East-European Conference, pp. 106–122. Tampere University of Technology, Pori (2008)

19. Kalinichenko, L.A., Stupnikov, S.A.: Heterogeneous information model unification as a pre-requisite to resource schema mapping. In: D'Atri, A., Saccà, D. (eds.) Information Systems: People, Organizations, Institutions, and Technologies - Proceedings of the V Conference of the Italian Chapter of Association for Information Systems itAIS, pp. 373–380. Springer Physica Verlag, Heidelberg (2010)

20. Kalinichenko, L.A., Stupnikov, S.A.: OWL as yet another data model to be integrated. In: Advances in Databases and Information Systems: Proceedings II of the 15th East-European Conference, pp. 178–189. Austrian Computer Society, Vienna (2011)

21. Kalinichenko, L., Stupnikov, S., Vovchenko, A., Kovalev, D.: Rule-based multi-dialect infrastructure for conceptual problem solving over heterogeneous distributed information resources. In: Catania, B., et al. (eds.) New Trends in Databases and Information Systems. Advances in Intelligent Systems and Computing, vol. 241, pp. 61–68. Springer, Cham (2014)

22. Kalinichenko, L.A., Stupnikov, S.A., Vovchenko, A.E., Kovalev, D.Y.: Conceptual modeling of multi-dialect workflows. Informatics and Applications **8**(4), 110–124 (2014)

23. Kopcke, H., Thor, A., Rahm, E.: Evaluation of entity resolution approaches on real-world match problems. Proc. VLDB Endowment **3**(1–2), 484–493 (2010)

24. Larsen, P.G., Plat, N., Toetenel, H.: A formal semantics of data flow diagrams. Formal Aspects Comput. **6**(6), 586–606 (1994)

25. Lano, K., Bicarregui, J., Evans, A.: Structured axiomatic semantics for UML models. In: Rigorous Object-Oriented Methods: Proceedings of the Conference, p. 5 (2000)

26. Lano, K., Kolahdouz-Rahimi, S., Clark, T.: Language-independent model transformation verification. In: Verification of Model Transformations, Proceedings of the Third International Workshop on Verification of Model Transformations, CEUR Workshop Proceedings, vol. 1325, pp. 36–45 (2014)

27. Miner, D.: MapReduce Design Patterns: Building Effective Algorithms and Analytics for Hadoop and Other Systems. O'Reilly Media, Sebastopol (2012)

28. Naumann, F., Bilke, A., Bleiholder, J., Weis, M.: Data fusion in three steps: resolving inconsistencies at schema-, tuple-, and value-level. IEEE Data Engineering Bulletin **29**(2), 21–31 (2006)

29. Olston, C., Reed, B., Srivastava, U., Kumar, R., Tomkins, A.: Pig Latin: a not-so-foreign language for data processing. In: Proceedings of the SIGMOD Conference, pp. 1099–1110 (2008)

30. Bellahsene, Z., Bonifati, A., Rahm, E. (eds.): Schema Matching and Mapping. Springer, Heidelberg (2011)

31. Stupnikov, S., Kalinichenko, L., Bressan, S.: Interactive discovery and composition of complex web services. In: Manolopoulos, Y., Pokorný, J., Sellis, T.K. (eds.) ADBIS 2006. LNCS, vol. 4152, pp. 216–231. Springer, Heidelberg (2006). doi:10.1007/11827252_18

32. Vassiliadis, P., Simitsis, A., Georgantas, P., Terrovitis, M., Skiadopoulos, S.: A generic and customizable framework for the design of ETL scenarios. Inf. Syst. **30**(7), 492–525 (2005)

33. Stupnikov, S.A.: Modeling of compositional refining specifications. Ph.D. thesis. Institute of Informatics Problems, Russian Academy of Sciences, Moscow, 195 p. (2006)

34. Stupnikov, S.A.: Unification of an array data model for the integration of heterogeneous information resources. In: Proceedings of the 14th Russian Conference on Digital Libraries RCDL 2012, CEUR Workshop Proceedings, vol. 934, pp. 42–52 (2012)

35. Stupnikov, S.A.: Mapping of a graph data model into an object-frame canonical information model for the development of heterogeneous information resources integration systems. In: Proceedings of the 15th Russian Conference on Digital Libraries RCDL 2013, CEUR Workshop Proceedings, vol. 1108, pp. 85–94 (2013)

36. Stupnikov, S.A., Vovchenko, A.E.: Combined virtual and materialized environment for integration of large heterogeneous data collections. In: Proceedings of the 16th Russian Conference on Digital Libraries RCDL 2014 Proceedings. CEUR Workshop Proceedings, vol. 1297, pp. 201–210 (2014)

37. InfoSphere Big Match for Hadoop. Technical Overview. https://goo.gl/0TMqvw

38. HIL2AMN Project. GitHub Repository (2017). https://goo.gl/IK1JzU

39. Stupnikov, S.: Formal semantics of a language for entity resolution and data fusion and its application for verification of data integration workflows. Selected Papers of the XVIII International Conference on Data Analytics and Management in Data Intensive Domains (DAMDID/RCDL 2016), CEUR Workshop Proceedings, vol. 1752, pp. 159–167 (2016)
40. Steinberg, D., Budinsky, F., Paternostro, M., Merks, E.: EMF: Eclipse Modeling Framework, 2nd edn. Addison-Wesley Professional, Reading (2008)
41. EMFText Concrete Syntax Mapper. http://www.emftext.org/index.php/EMFText

A Study of Several Matrix-Clustering Vertical Partitioning Algorithms in a Disk-Based Environment

Viacheslav Galaktionov[1], George Chernishev[1,2(✉)], Kirill Smirnov[1], Boris Novikov[1], and Dmitry A. Grigoriev[1]

[1] Saint-Petersburg State University, Saint-Petersburg, Russia
viacheslav.galaktionov@gmail.com,
{g.chernyshev,k.k.smirnov,b.novikov,d.a.grigoriev}@spbu.ru
[2] JetBrains Research, Saint-Petersburg, Russia

Abstract. In this paper we continue our efforts to evaluate matrix clustering algorithms. In our previous study we presented a test environment and results of preliminary experiments with the "separate" strategy for vertical partitioning. This strategy assigns a separate vertical partition for every cluster found by the algorithm, including inter-submatrix attribute group. In this paper we introduce two other strategies: the "replicate" strategy, which replicates inter-submatrix attributes to every cluster and the "retain" strategy, which assigns inter-submatrix attributes to their original clusters. We experimentally evaluate all strategies in a disk-based environment using the standard TPC-H workload and the PostgreSQL DBMS. We start with the study of record reconstruction methods in the PostgreSQL DBMS. Then, we apply partitioning strategies to three matrix clustering algorithms and evaluate both query performance and storage overhead of the resulting partitions. Finally, we compare the resulting partitioning schemes with the ideal partitioning scenario.

Keywords: Database tuning · Vertical partitioning · Experimentation · Matrix clustering · Fragmentation · TPC-H · PostgreSQL

1 Introduction

The vertical partitioning problem [4] is one of the oldest problems in the database domain. There are dozens or even hundreds of studies available on the subject. It is a subproblem of the general database physical structure selection problem. It can be described as follows [8]: find a configuration (a set of vertical fragments) which would satisfy the given constraints and provide the best performance. There are two major classes of approaches to this problem:

- Cost-based approach [2, 15, 20, 32]. Studies that follow this approach construct a cost model which is used to predict the performance of a workload for any given configuration. Next, an algorithm enumerating the configuration space is used.
- Procedural approach [28, 31, 34]. These studies do not use the notion of configuration cost. Instead, they propose some kind of a procedure which will result in a "good"

© Springer International Publishing AG 2017
L. Kalinichenko et al. (Eds.): DAMDID/RCDL 2016, CCIS 706, pp. 163–177, 2017.
DOI: 10.1007/978-3-319-57135-5_12

configuration. Usually, these studies provide some intuitive explanation why the ensuing configuration would be "good".

The abundance of studies is justified by the following considerations:

- It was proved that the problem of vertical partitioning is an NP-hard problem [3, 28, 36], just like many other physical design problems [5, 21, 36].
- Estimation errors related to both the system parameters and workload parameters. System parameters (hardware and software) in some cases cannot be measured precisely. Workload parameters can also be imprecise, e.g. not all queries are known in advance, or some of the known queries are not run. All these errors can cause the performance of the solution to deteriorate.

The procedural approach was very popular in the '80s and '90s due to the lack of computational resources. Nowadays, the interest for it has largely declined, and the majority of contemporary studies follows the cost-based one. This approach produces more accurate recommendations by incorporating additional information into the selection process. However, procedural approach has a number of promising applications:

- Dynamization of vertical partitioning [23, 27, 33, 35]. All of the previous vertical partitioning studies considered the problem in a static context, i.e. a configuration is selected once. In case of changes in the workload or the data the algorithm has to be re-run. In the new formulation the goal is to adapt the partitioning scheme to a constantly changing workload. The straightforward technique of the repeated re-run of a cost-based algorithm is not applicable due to its formidable costs of operation. Otherwise, its application will result in query processing stalls which should be avoided at all costs in this formulation. However, the procedural approach is not so computationally demanding as the cost-based one. Thus, low-quality solutions are acceptable as long as they provide improvement over the previous configuration and help us avoid query processing stalls.
- Big data applications or any other cases featuring constrained resources.
- Tuning of multistores [26] or any other case when no details or only inaccurate estimates of physical parameters are known. It was already noted in the '80s [30] that the procedural approach is well-suited for such cases. A multistore system is a database system which consists of several distinct data stores, e.g. a Hadoop HDFS and an RDBMS. This kind of a system is a modern example of the case where not every physical parameter of underlying data stores is known.

This paper is an extended version of the paper [17]. In our previous study we presented a test environment and results of preliminary experiments with "separate" strategy for vertical partitioning. This strategy assigns a separate vertical partition for every cluster found by the algorithm, including the inter-submatrix attribute group. The preliminary results showed little to no performance improvement with this strategy for all three algorithms for in-memory environment.

In this paper we continue our study of matrix clustering algorithms. This time we consider a disk-based environment. Firstly, we try to improve record reconstruction times via several approaches. Next, we introduce two strategies:

- A "replicate" strategy, which replicates inter-submatrix attributes to every cluster.
- A "retain" strategy, which assigns inter-submatrix attributes to their original clusters.

We study these strategies and compare them with the "separate" strategy and the unpartitioned case. We evaluate all strategies in terms of query performance and storage overhead. Finally, we compare the resulting partitioning schemes with the ideal partitioning scenario, where each query gets a specially-tuned fragment.

2   Related Work

2.1   Classification

The vertical partitioning problem is one of the oldest problems in the database domain. There are several dozens of studies on this topic, and most of them concern various algorithms. Several surveys can be found in [13, 14]. Vertical partitioning algorithms can be classified into two major groups: cost-based and procedural, where the latter employs three types of approaches:

- Attribute affinity and matrix clustering approaches [9, 10, 12, 18, 22]. In affinity-based approaches, closeness between every two attributes is first calculated, and then it is used to define the borders of the resulting fragments. This closeness is called attribute affinity. At the first step a workload is used to create an AUM, then an Attribute Affinity Matrix (AAM) is constructed using a paper-specific transformation procedure. Finally, a row and column permutation algorithm is applied. Matrix clustering approaches operate on the AUM and start with the permutation part.
- Graph approaches [11, 16, 28, 31, 38]. Most of the graph approaches treat the AAM as an adjacency matrix of an undirected weighted graph. In this graph nodes denote attributes and edges represent a bounds strength. Then a template is sought by various means, e.g. kruskal-like algorithms or hamiltonian way cut. The resulting templates are used to construct partitions.
- Data mining approaches [7, 19, 34]. This is a relatively new vertical partitioning technique that uses association rules to derive vertical fragments. Most of these works mine a workload (a transaction set) for rules which use sets of attributes as items. In these studies, existing algorithms for association rule search are used to uncover relations between attributes. In particular, an adapted Apriori [1] algorithm is a very popular choice.

Let us review the matrix clustering approach in detail.

2.2   Matrix Clustering Approach

The general scheme of this approach is the following:

- Construct an Attribute Usage Matrix (AUM) from the workload. The matrix is constructed as follows:

$$M_{ij} = \begin{cases} 1, & query\ i\ uses\ attribute\ j \\ 0, & otherwise \end{cases}$$

- Cluster the AUM by permuting its rows and columns to obtain a block diagonal matrix.
- Extract these blocks and use them to define the resulting partitions.

Some approaches do not operate on a 0–1 matrix. Instead they modify matrix values to account for additional information like query frequency, attribute size and so on. Let us consider an example. Suppose that we have six queries accessing six attributes:

The next step is the creation of an AUM using this workload. The resulting matrix is shown in Fig. 1a. Having applied a matrix clustering algorithm, we acquire the reordered AUM (Fig. 1b). The resulting fragments are the following: (a, b), (b, f), (d, e).

```
q1: SELECT a FROM T WHERE a > 10;
q2: SELECT b, f FROM T;
q3: SELECT a, c FROM T WHERE a = c;
q4: SELECT a FROM T WHERE a < 10;
q5: SELECT e FROM T;
q6: SELECT d, e FROM T WHERE d + e > 0;
```

	a	b	c	d	e	f
q1	1	0	0	0	0	0
q2	0	1	0	0	0	1
q3	1	0	1	0	0	0
q4	1	0	0	0	0	0
q5	0	0	0	0	1	0
q6	0	0	0	1	1	0

(a) AUM

	a	c	b	f	d	e
q1	1	0	0	0	0	0
q3	1	1	0	0	0	0
q4	1	0	0	0	0	0
q2	0	0	1	1	0	0
q6	0	0	0	0	1	1
q5	0	0	0	0	0	1

(b) Reordered AUM

Fig. 1. Matrix clustering algorithm

However, not all matrices are fully decomposable. Consider the matrix presented in Fig. 2. The first column obstructs the perfect decomposition into several clusters. In this case, the algorithm should produce a decomposition which would minimally harm query processing and would result in an overall performance improvement. Matrix clustering algorithms employ different strategies to select such a decomposition.

a	b	c	d	e	f
1	1	1	0	0	0
1	1	1	0	0	0
1	1	1	0	0	0
1	0	0	1	1	0
1	0	0	1	1	0
1	0	0	0	0	1

Fig. 2. Non-decomposable matrix

2.3   Matrix Clustering Algorithms

The first study to introduce matrix clustering to vertical partitioning was the work of Hoffer [22]. The idea is to store together (in one file) attributes possessing identical retrieval patterns. The patterns are expressed through the notion of attribute cohesion, which shows how attributes in a pair are related to each other. The author proposes a pairwise attribute similarity measure to capture this cohesion.

The proposed measure relies on three parameters: co-access frequency of a pair of attributes, attribute length and relative importance of the query. This measure was designed having the following properties in mind: it is non-decreasing by co-access frequency, non-decreasing by both attribute lengths (individually) and the function is non-increasing in the combined length of attributes.

Finally, having an attribute affinity matrix, an existing clustering algorithm (Bond Energy Algorithm, BEA) [29] is used. It permutes rows and columns to maximize nearest neighbor bond strengths. The author was motivated in his choice by the following: this algorithm is insensitive to the order in which items are presented; it has a low computation time, etc. However, this algorithm has a disadvantage: it requires human attention for cluster selection.

BEA is not the only existing matrix clustering algorithm. Another permutation algorithm was proposed in the reference [37]. Similarly to BEA, it permutes rows and columns, but tries to minimize the spanning path of the graph represented by the original matrix. The improvement of these two algorithms is presented in the reference [6]. This algorithm is called the matching algorithm and it uses Hamming distance to produce clusters. According to [9], the study [24] presents the Rank Order algorithm. Its idea is to sort rows and columns of the original matrix in descending order of their binary weight. The Cluster Identification (CI) algorithm by Kusiak and Chow [25] is an algorithm for clustering 0–1 matrices. The proposed approach is to detect clusters one by one using a special procedure. This procedure resembles the search of a transitive closure for rows and columns. It is an optimal algorithm that can solve the problem when the matrix is perfectly separable, e.g. when clusters do not intersect (there is no attribute sharing).

All of the aforementioned algorithms (except BEA) are generic matrix clustering algorithms. They do not address the vertical partitioning problem and do not even bear any database specifics. The next studies by Chun-Hung Cheng [9, 10, 12] attempt to apply matrix clustering approach to the database domain. Several new vertical partitioning algorithms were developed in his works. Let us consider them.

Chun-Hung Cheng criticizes existing matrix clustering algorithms [9, 10]:

- They do not always produce a solution matrix in a diagonal submatrix structure. Thus, these algorithms may require additional computation to extract them;
- These algorithms may require decision of database administrator to identify inter-submatrix attributes [9].

The first study [9] extends the original CI [25] algorithm to non-decomposable matrices. The proposed approach is to remove columns obstructing the decomposition (inter-submatrix attributes).

The author considered the following problem formulation $P1$ [9]: remove columns to decompose a matrix into separable submatrices with the maximum number of "1" entries retained in submatrices subject to the following constraints:

- C1: A submatrix must contain at least one row;
- C2: The number of rows in a submatrix cannot exceed upper limit, b;
- C3: A submatrix must contain at least one column.

To solve the problem, the branch and bound approach was used. This approach uses an objective function which maximizes the number of "1" entries in the resulting submatrices. During the tree traversal, upper and lower bounds are calculated and used to guide the enumeration process.

However, the basic approach required traversal of too many nodes, so the author augmented it with the following heuristic. A so-called **blocking measure** is calculated for each column. It estimates the likelihood of a column being an obstacle to the further decomposition of the matrix. Basically, it is the number of columns that would be involved in all queries which use the given attribute. Next, the columns are ordered by their respective values and the ones with the highest values are checked.

The study [10] also extends the original CI algorithm. The author adopts the same branch and bound approach as in his previous paper [9]. However, instead of the **blocking measure** a new **void measure** is developed. It has the same purpose, which is the estimation of the likelihood of a column being an inter-submatrix column. Essentially, this measure is the calculated "free space" to the left and to the right of the candidate cluster.

The next study of the author [12] addresses several shortcomings of his previous works:

- The problem of the parameter b. While this parameter helps prevent the formation of the huge clusters, it does not guarantee any quality of the resulting clusters. Also, the problem will have to be reformulated if several clusters of different sizes are needed.
- The dangling transaction problem. Applying the previous algorithm [10] a transaction not belonging to any cluster may be acquired: all of its attributes would be removed. Two examples are presented in the original paper.
- The previous work did not include such an important parameter as the access frequency of the transactions.

Thus, a new formulation $P3$ is proposed [12]: remove a minimal number of "1" entries to decompose a transaction-attribute access matrix into separable submatrices subject to the following constraints:

- C7: Transactions with all "0" entries in a submatrix are not allowed.
- C8: Attributes with all "0" entries in a submatrix are not allowed.
- C9: The **cohesion measure** of a submatrix is more than or equal to a threshold, δ.

Cohesion measure of a submatrix is the ratio of "1" elements to "0" elements. This new measure is used to ensure the quality of a cluster.

The problem is also solved with the branch and bound approach, again, the **void measure** is used to guide the order of node traversal.

Furthermore, in this work the author shows why dangling transactions should be avoided: an example is provided showing a case where it is possible to lose information regarding a cluster. Finally, the author extended his CI framework to consider query frequencies. This $P4$ formulation is the same as $P3$, but features a weighted sum of accesses [12]: minimize the loss of total accesses $\left(\sum_i \sum_j a_{ij} * freq_i \right)$ due to the removal of a_{ij} for decomposing a transaction-attribute matrix into separable submatrices subject to the same constraints C7–C9. In this paper we study the approaches described in the references [9, 10, 12].

3 System Architecture

We have developed a program for experimental evaluation of the considered algorithms. Its architecture is presented on Fig. 3. It consists of the following modules:

- The parser reads the workload from a file. It extracts the queries and passes them to the executor, so that their execution times can be measured. It also constructs the AUM, which serves as input for the selected algorithm.
- The algorithm identifies clusters and passes that information to the partitioner to create corresponding temporary tables.
- The query rewriter also receives this information. It replaces the name of the original table with the ones that were generated by the partitioner. It can handle subqueries; view support is not implemented yet.
- The partitioner generates new names and sends partitioning commands to the database. The exact commands are SELECT INTO and ALTER TABLE. The latter lets it transfer primary keys.
- The executor accepts queries and sends them to PostgreSQL to measure the time of execution.

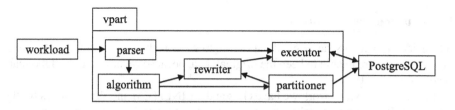

Fig. 3. The architecture of our approach

4   Experiments

4.1   A Brief Summary of Previous Findings

In our previous paper [17] we have implemented three recent matrix clustering algorithms [9, 10, 12] (A94, A95, A09) and used PostgreSQL DBMS to evaluate them. Our experiments were conducted using the standard benchmark—TPC-H with scale factor 1. The database was placed in the main memory of the machine. To accomplish this, the PostgreSQL data directory was put on a tmpfs partition, created with standard GNU/Linux utilities. We have employed a "separate" strategy for vertical partitioning. This strategy assigns a separate vertical partition for every cluster found by the algorithm, including inter-submatrix attribute group.

Experiments showed that all of the considered algorithms perform poorly in this environment, often yielding partitioning schemes worse than the original one.

4.2   Experimental Plan

In this paper we try to analyze the reasons for this outcome. The plan of this study is the following:

1. First of all, we changed the environment from the main memory to the disk-based one. It is used for all of the experiments described in this paper.
2. Then, we performed the evaluation of record reconstruction methods of PostgreSQL. We try different methods to speed-up or to completely avoid record reconstruction expenses. Eventually, we select the best record reconstruction method and use it throughout the rest of the paper.
3. Next, we introduce two new partition generation strategies called "replicate" and "retain". We compare them with each other, with the original non-partitioned configuration and the "separate" strategy employed in our previous work.
4. Finally, we compare our strategies with the ideal partitioning scenario, where each query gets a specially-tuned fragment.

4.3   Data and Query Setup

We decided to stick to the original data setup to support the reproducibility of results. However, we were unable to keep the original data, because we have changed the scale factor from 1 to 10.

Like in our previous work we have chosen the LINEITEM and PART tables for the evaluation. Based on these tables we have formulated the following query setups:

– Query Setup 1 (QS1): Q1, Q6, Q14, Q19;
– Query Setup 2 (QS2): Q6, Q14, Q19;
– Query Setup 3 (QS3): Q6, Q14.

It is important to note that only queries Q14 and Q19 involve join of the LINEITEM and PART tables. Queries Q1 and Q6 contain only filtering predicates and aggregate functions on the LINEITEM table.

4.4 Hardware and Software Setups, Measurements

In our experiments we have used the following setup (almost the same as in our previous study):

– PostgreSQL 9.6.1,
– Funtoo Linux (kernel 4.8.4),
– Intel Core i7-3630QM (4 physical cores, hyper-threading enabled),
– 8 GB (DDR3) RAM,
– GCC 4.9.3.

All numbers presented in this paper are averages over five runs. During evaluation we noticed hot cache effects: the difference between the first and the fifth run may be about 3–5 times. Thus, to negate these effects we restarted PostgreSQL and dropped OS caches for each query. This way we obtained stable, almost equal to each other results for each value in a series of measurements.

4.5 Disk-Based Setup Re-run

All six scenarios from our previous study were re-run in the disk-based environment. It is important to note that we benchmarked the same partitioning schemes, because the algorithm does not depend on the environment-related (physical) parameters.

The benchmarking results were essentially the same, the partitioned scheme was worse than the non-partitioned one in terms of query performance. The overall performance of a query batch degraded 2–3 times, depending on the scenario. Due to the space constraints we will not present all of the scenarios. However, we will closely study scenario 1 in the "New Fragmentation Strategies" section.

4.6 Record Reconstruction Study

During the re-run we found out that the major performance hit was the join cost. Thus, we tried to improve the efficiency of record reconstruction. We have considered the following techniques:

1. Default, where we just rewrite queries with joins.
2. Sequential scan, where we forced sequential scan for reading table data.
3. Index scan, where we made PostgreSQL use the index on join attributes for query processing.

For our experiments we used only query Q1, because it does not have any joins with the PART table in its execution plan. This way we will study the effects of just the reconstruction joins. The results of our experiments are presented in Fig. 10. Here you can see that default and sequential scan techniques are indistinguishable from each other. On the other hand, index scan performance is significantly worse than both of them. We examined the query plan and noticed that the performance deterioration presents in both scan operators and in the join operator. We also tried to use a clustered index, but performance was worse than with an index scan. Thus, we decided to stick to sequential scan reconstruction technique throughout the paper. Also, note the consistent behavior of these techniques for the "replicate" strategy (we will describe it in the next subsection).

4.7 New Fragmentation Strategies

In our previous paper we studied the "separate" strategy, which assigns a separate vertical partition for every cluster found by the algorithm, including inter-submatrix attribute group. The author of these matrix clustering algorithms indicated several other possible strategies to form vertical partitions. In this paper we are interested in the following:

- A "replicate" strategy, which replicates inter-submatrix attributes to every cluster. This strategy should eliminate all joins by introducing storage overhead. It should also degrade the performance of insert and update queries, because of the need to keep data synchronized.
- A "retain" strategy, which assigns inter-submatrix attributes to their original clusters. This strategy is inherent only to algorithm A09, which removes accesses in the matrix. The columns, access to which has been removed, are considered intersubmatrix and must be handled appropriately. However, the clustering algorithm itself already returns them as parts of some clusters. The point of this strategy is to keep them in place.

In Fig. 4 you can see the performance of the A09 algorithm in the disk-based environment using strategies "separate", "replicate" and "retain" for scenario 1. These graphs are given along with the "original" (unpartitioned) variant. The corresponding partitioning schemes can be found in our previous study.

The general conclusions for this experiment are the following: the configuration generated by the "retain" strategy shows the worst performance, the configuration generated by the "separate" strategy is still worse than the unpartitioned data. On the other hand, the "replicate" strategy's configuration performs significantly better than the unpartitioned one. This behavior is explained by the record reconstruction expenses which have to be performed using a relational join operation. These expenses significantly outweigh the benefits of the vertical partitioning.

However, the "replicate" strategy does not come without a price. Replication of attributes increases the amount of data that needs to be stored. Thus, we decided to assess storage overhead. Figure 5 presents the respective table sizes for all strategies, alongside with the non-partitioned configuration for comparison. Each section of a bar column marks the size of the LINEITEM table partition. Thus, "separate" strategy partitions data into three pieces, while "retain" partitions into two. From this experiment we can see that the "retain" and "replicate" strategies produce very little overhead, while "separate" strategy requires almost 1.5 times more space than the original scheme. Such difference can be justified by the need to copy primary key to every partition (2 vs 3) and the fact that in LINEORDER the table primary key is a multi-attribute entity, which is rather big. Probably, we could reduce the amount of required space by introducing our own (smaller) primary key, solely for reconstruction.

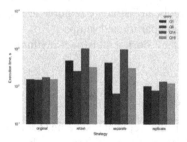

Fig. 4. Strategies comparison, scenario 1

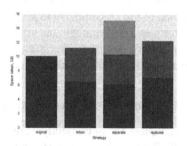

Fig. 5. Disk requirements, scenario 1

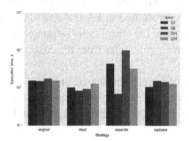

Fig. 6. Strategies comparison, scenario 6

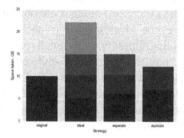

Fig. 7. Disk requirements, scenario 6

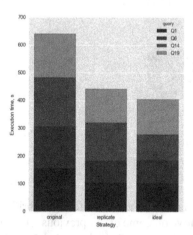

Fig. 8. Ideal Performance

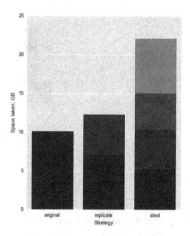

Fig. 9. Ideal disk usage

Figure 6 contains the performance of the A95 algorithm in a disk-based environment using strategies "separate" and "replicate" for scenario 6 (the strategy "ideal" is discussed in the next section). These graphs are given alongside with the original variant, for the ease of comparison. Here we can also see that the "separate" strategy performs worse than the unpartitioned variant, like in the in-memory case. The "replicate" strategy is the winner in this scenario as well. The "retain" strategy is inapplicable to the A95 algorithm, so it is absent in this graph. Disk space requirements for this scenario are presented in Fig. 7. Similarly to scenario 1, the "replicate" strategy requires less space than the "separate" strategy, but still more than the unpartitioned setup.

We do not describe other scenarios because they do not illustrate anything new compared to the ones considered in this subsection.

4.8   Comparison with Ideal Partitioning Scheme

The next questions of our study were "How good is the 'replicate' scheme? Can we do better?". To answer them we have introduced an ideal partitioning scenario, where each query gets a fragment containing the minimum number of attributes.

We had re-run scenario 1 (algorithm A09) and compared the "replicate" strategy with the ideal partitioning scenario. The results are presented in Figs. 8 and 9. As you can see, the ideal scheme is about 10% better than the "replicate" strategy while consuming almost twice as much space. This improvement comes from the costs of Q14. Thus, "replicate" strategy is quite efficient.

There is a similar picture with scenario 6 (algorithm A95) presented in Figs. 6 and 7. This time the ideal scheme is 24% better than the "replicate" strategy with the same storage overhead. The major sources of improvement are queries Q6 (almost two times) and Q14 (about 1.5 times).

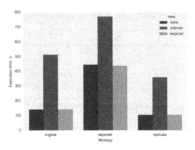

Fig. 10.   Record reconstruction

4.9   Conclusions

This is the continuation of our previous study [17] of the matrix clustering algorithms. In this paper we analyze the reasons for the negative outcome of our previous work and explore alternative vertical partition generation strategies. To achieve better understanding of the subject we switched to disk-based configuration, as opposed to in-memory one used in our previous work.

Our experiments showed that PostgreSQL encounters performance problems while reconstructing vertically partitioned records. We had to use relational join operation and we were unable to speed-up it by using regular and clustered indexes.

Next, we found out that the "separate" strategy, which was the only one strategy used in our previous study and which produced partitions that were consistently worse than the unpartitioned schema, behaved the same way in the disk-based environment. Moreover, we demonstrated that in terms of disk space required for the partitioning, this strategy is also very expensive.

On the other hand, the performance of the schemes generated by the new strategy "replicate" were significantly (20%–50% depending on a case) better than the unpartitioned schema. Storage overhead was about 20% that of the unpartitioned case. The other strategy – "retain" generated schemes which performed worse than generated by the "separate" strategy (thus being the worst strategy), while requiring only 10% more storage than in unpartitioned case.

We also compared the "replicate" strategy with the ideal partitioning scheme, where each query receives a specially-tailored minimal partition. To our surprise, the quality of "replicate" strategy schemes was quite close to the ideal one. Meanwhile, the storage overhead of the ideal scheme was almost 2.5 times that of the unpartitioned one and the one generated by "replicate" was only 1.2.

We can conclude with the following:

- The "replicate" strategy generates near ideal schemes, but with the reasonable storage overhead. These schemes provide significant improvement over unpartitioned case for read-only workloads.
- Implementing vertical partitioning one should avoid record reconstruction or should try to implement it without using costly relational join operation.

References

1. Agrawal, R., Srikant, R.: Fast algorithms for mining association rules in large databases. In: VLDB 1994 Conference Proceedings, pp. 487–499 (1994)
2. Agrawal, S., Narasayya, V., Yang, B.: Integrating vertical and horizontal partitioning into automated physical database design. In: SIGMOD 2004 Conference Proceedings, pp. 359–370 (2004). doi:10.1145/1007568.1007609
3. Apers, P.M.G.: Data allocation in distributed database systems. ACM Trans. Database Syst. **13**, 263–304 (1988). doi:10.1145/44498.45063
4. Bellatreche, L.: Optimization and tuning in data warehouses. In: Liu, L., Özsu, M.T. (eds.) Encyclopedia of Database Systems, pp. 1995–2003. Springer, New York (2009). doi:10.1007/978-0-387-39940-9_259
5. Bellatreche, L., Boukhalfa, K., Richard, P.: Data partitioning in data warehouses: hardness study, heuristics and Oracle validation. In: DAWAK 2008 Conference Proceedings, pp. 87–96 (2008). doi:10.1007/978-3-540-85836-2_9
6. Bhat, M.V., Haupt, A.: An efficient clustering algorithm. IEEE Trans. Syst. Man Cybern. **6**(1), 61–64 (1976). doi:10.1109/TSMC.1976.5408399
7. Bouakkaz, M., Ouinten, Y., Ziani, B.: Vertical fragmentation of data warehouses using the FP-Max algorithm. In: IIT 2012 Conference Proceedings, pp. 273–276 (2012)

8. Chaudhuri, S., Weikum, G.: Self-management technology in databases. In: Liu, L., Özsu, M.T. (eds.) Encyclopedia of Database Systems, pp. 2550–2555. Springer, New York (2009). doi:10.1007/978-0-387-39940-9_334

9. Cheng, C.: Algorithms for vertical partitioning in database physical design. Omega **22**(3), 291–303 (1994). doi:10.1016/0305-0483(94)90042-6

10. Cheng, C.-H.: A branch and bound clustering algorithm. IEEE Trans. Syst. Man Cybern. **25**(5), 895–898 (1995). doi:10.1109/21.376504

11. Cheng, C.-H., Lee, W.-K., Wong, K.-F.: A genetic algorithm-based clustering approach for database partitioning. IEEE Trans. Syst. Man Cybern. Part C Appl. Rev. **32**(3), 215–230 (2002). doi:10.1109/TSMCC.2002.804444

12. Cheng, C.-H., Motwani, J.: An examination of cluster identification-based algorithms for vertical partitions. Int. J. Bus. Inf. Syst. **4**(6), 622–638 (2009). doi:10.1504/IJBIS.2009.026695

13. Chernishev, G.: Towards self-management in a distributed column-store system. In: Morzy, T., Valduriez, P., Bellatreche, L. (eds.) ADBIS 2015. CCIS, vol. 539, pp. 97–107. Springer, Cham (2015). doi:10.1007/978-3-319-23201-0_12

14. Chernishev, G.: Vertical partitioning in relational DBMS. In: Talk at the Moscow ACM SIGMOD chapter meeting; slides and video. http://synthesis.ipi.ac.ru/sigmod/seminar/s20150430

15. Chu, W., Ieong, I.: A transaction-based approach to vertical partitioning for relational database systems. IEEE Trans. Softw. Eng. **19**(8), 804–812 (1993)

16. Du, J., Barker, K., Alhajj, R.: Attraction - a global affinity measure for database vertical partitioning. In: IADIS ICWI 2003 Conference Proceedings, pp. 538–548 (2003)

17. Galaktionov, V., Chernishev, G., Novikov, B., Grigoriev, D.: Matrix clustering algorithms for vertical partitioning problem: an initial performance study. In: Selected Papers of the XVIII International Conference on Data Analytics & Management in Data Intensive Domains (DAMDID/RCDL 2016), Ershovo, Moscow Region, Russia, CEUR Workshop Proceedings, vol. 1752, pp. 24–31 (2016)

18. Gorla, N., Boe, W.J.: Database operating efficiency in fragmented databases in mainframe, mini, and micro system environments. Data Knowl. Eng. **5**(1), 1–19 (1990). doi:10.1016/0169-023X(90)90030-H

19. Gorla, N., Yan, B.P.W.: Vertical fragmentation in databases using data-mining technique. In: Erickson, J. (ed.) Database Technologies: Concepts, Methodologies, Tools, & Applications, pp. 2543–2563. IGI Global (2009)

20. Hammer, M., Niamir, B.: A heuristic approach to attribute partitioning. In: SIGMOD 1979 Conference Proceedings, pp. 93–101 (1979). doi:10.1145/582095.582110

21. Harinarayan, V., Rajaraman, A., Ullman, J.D.: Implementing data cubes efficiently. ACM SIGMOD Record **25**(2), 205–216 (1996). doi:10.1145/235968.233333

22. Hoffer, J.A., Severance, D.G.: The use of cluster analysis in physical data base design. In: VLDB 1975 Conference Proceedings, pp. 69–86 (1975). doi:10.1145/1282480.1282486

23. Jindal, A., Dittrich, J.: Relax and let the database do the partitioning online. In: BIRTE 2011 Workshop Proceedings, pp. 65–80 (2012). doi:10.1007/978-3-642-33500-6_5

24. King, J.R.: Machine-component grouping in production flow analysis: an approach using a rank order clustering algorithm. Int. J. Prod. Res. **18**(2), 213–232 (1980)

25. Kusiak, A., Chow, W.: An efficient cluster identification algorithm. IEEE Trans. Syst. Man Cybern. **17**(4), 696–699 (1987)

26. LeFevre, J., Sankaranarayanan, J., Hacigumus, H., Tatemura, J., Polyzotis, N., Carey, M.J.: MISO: souping up big data query processing with a multistore system. In: SIGMOD 2014 Conference Proceedings, pp. 1591–1602 (2014). doi:10.1145/2588555.2588568

27. Li, L., Gruenwald, L.: Self-managing online partitioner for databases (SMOPD): a vertical database partitioning system with a fully automatic online approach. In: IDEAS 2013 Conference Proceedings, pp. 168–173 (2013). doi:10.1145/2513591.2513649

28. Lin, X., Orlowska, M., Zhang, Y.: A graph based cluster approach for vertical partitioning in database design. Data Knowl. Eng. **11**(2), 151–169 (1993)

29. McCormick, W., Schweitzer, P., White, W.: Problem decomposition and data reorganization by a clustering technique. Oper. Res. **20**(5), 993–1009 (1972)

30. Navathe, S., Ceri, S., Wiederhold, G., Dou, J.: Vertical partitioning algorithms for database design. ACM Trans. Database Syst. **9**, 680–710 (1984)

31. Navathe, S.B., Ra, M.: Vertical partitioning for database design: a graphical algorithm. In: SIGMOD 1989 Conference Proceedings, pp. 440–450 (1989). doi:10.1145/67544.66966

32. Papadomanolakis, S., Ailamaki, A.: An integer linear programming approach to database design. In: ICDE 2007 Workshop Proceedings, pp. 442–449 (2007)

33. Rodríguez, L., Li, X.: A dynamic vertical partitioning approach for distributed database system. In: IEEE SMC Conference Proceedings, pp. 1853–1858 (2011). doi:10.1109/ICSMC.2011.6083941

34. Rodríguez, L., Li, X.: A support-based vertical partitioning method for database design. In: CCE 2011 Conference Proceedings, pp. 1–6 (2011). doi:10.1109/ICEEE.2011.6106682

35. Rodríguez, L., Li, X., Mejía-Alvarez, P.: An active system for dynamic vertical partitioning of relational databases. In: MICAI 2011 Conference Proceedings, pp. 273–284 (2011). doi:10.1007/978-3-642-25330-0_24

36. Sacca, D., Wiederhold, G.: Database partitioning in a cluster of processors. ACM Trans. Database Syst. **10**, 29–56 (1985). doi:10.1145/3148.3161

37. Slagle, J.R., Chang, C.L., Heller, S.R.: A clustering and data-reorganizing algorithm. IEEE Trans. Syst. Man Cybern. **5**(1), 125–128 (1975)

38. HyunSon, J., HoKim, M.: α-Partitioning algorithm: vertical partitioning based on the fuzzy graph. In: Mayr, H.C., Lazansky, J., Quirchmayr, G., Vogel, P. (eds.) DEXA 2001. LNCS, vol. 2113, pp. 537–546. Springer, Heidelberg (2001). doi:10.1007/3-540-44759-8_53

Clustering of Goods and User Profiles for Personalizing in E-commerce Recommender Systems Based on Real Implicit Data

Victor N. Zakharov[1(✉)] and Stanislav A. Philippov[2]

[1] Federal Research Center "Computer Science and Control", Russian Academy of Sciences,
Moscow, Russia
VZakharov@ipiran.ru
[2] National Research Nuclear University MEPhI, Moscow, Russia
stanislav@philippov.ru

Abstract. The work is devoted to description of a hybrid approach to the preparation of data for electronic commerce recommender systems. The efficiency of recommendations is improved via use of various algorithms depending on what is known about the user, little or no information or it is not the first visit and the browsing history is available. In the first case our own Item-Item CF method is used enabling to solve the problem of cold start. In the second case our own method User-User CF is applied. Both methods are based on clustering of both explicit and implicit user data. This approach can increase the number of items in the basket in fewer clicks in comparison with the methods that do not use the implicit data. The approach applicability was confirmed via test on real data obtained from Thaisoap, an online store.

Keywords: Pertinence · Collaborative filtering · E-commerce recommender system · Implicit data targeting · Item-Item CF · User-User CF · Cold start

1 Introduction

Even at the dawn of e-commerce in 1998, Amazon.com founder Jeff Bezos announced the need to personalize the information proposal with the phrase "If we have 4.5 million customers … we should have 4.5 million stores" [1]. The main practical objectives of personalization are:

- Providing relevant content. Personalization of site content simplifies the process of finding the necessary information for users and improves the overall efficiency of sites.
- Targeted advertising or targeting implies advertising centered around preferences of specific groups based on their interaction history with the site. There are many different forms of targeted advertising, the most promising being behavioral targeting (promotional offers based on user behavioral profile).

© Springer International Publishing AG 2017
L. Kalinichenko et al. (Eds.): DAMDID/RCDL 2016, CCIS 706, pp. 178–191, 2017.
DOI: 10.1007/978-3-319-57135-5_13

1.1   Personalization Methods

In general all personalization methods can be divided into two large groups: rule-based personalization and personalization based on algorithms. Personalization of sites based on the rules is using historical, geographic, demographic, and other data to generate proposals based on predefined rules. A typical personalization based on rules is "if the user does X, then show to him offer Y". For example, if the user of the site enters the section of used cars, he is invited to visit the sites of service stations (or auto parts stores) situated at a distance of not more than 50 km from the user's location. Rule-based personalization is applicable when users can be divided (segmented) into groups with distinctive features (for example, specific car brands owners' communities). Usually special services (e.g. Airee.co or Monoloop.com) are used for user segmentation and targeted offer development.

Personalization based on algorithms uses predicative models taking into account different kinds of information about user activity in order to create content fully meeting user's expectations. This approach is now considered to be the most effective; however it requires significant computing resources for storing behavioral data, its processing and analyzing as well as for defining recommendations. Such algorithms form the basis of recommender systems analyzing users' behavioral habits or estimating relevancy of goods based on their popularity among users in order to generate recommendations.

It is believed that providing personalized content to users can significantly increase the effectiveness of site, which is known in marketing as conversion rate (the number of visitors who have taken useful actions as defined by system owner divided by the total number of website visitors, in percent). For high-quality personalization of sites working with big number of users a complex approach combining marketing research and specific visitors' behavioral analysis is used. The information about marketing preferences of visitors can be also obtained via using web analytics systems such as Adobe Digital Marketing Suite or Google Analytics with Siteapps.com. In e-commerce the main tool of content personalization are recommender systems providing automatic processing of user activity data and recommending products and services that may be of interest to specific users [2]. The initial data for the analysis of user behavior is information about their activity, collected either explicitly or implicitly. Explicit personal data is received during registration, purchases, items rankings and reviews, as a result of polls and surveys, and on the base of other decisions of the user. Meanwhile the main information about user activity is collected implicitly by logging their actions, i.e. decision-making process. The results can be recorded in the log files or specialized databases. The subject of monitoring are the links clicked, time spent on separate pages, facts of the purchase of goods and services, geographical location of user etc. It should be noted that it is a vast array of data that is heterogeneous and rather difficult to interpret.

1.2   Defining User Preferences

Due to the large volume and heterogeneity of user activity data, methods of Data Mining are often used for its analyzing. These methods are intended to discover hidden (or previously unknown) knowledge in the investigated data sets (usually poorly structured).

Knowledge gained through the application of data mining techniques is usually presented in the form of regularities (patterns). They can look as association rules, decision trees, clusters, mathematical functions. The main tasks of Data Mining are classification, clustering, association, prediction and visualization. One of the most rapidly developing areas in Data Mining is the analysis of the relationships between the data (link analysis) applied in bioinformatics, digital libraries, and protection against terrorism [3]. All methods of data analysis used in Data Mining can be divided into two big groups: statistical methods and cybernetic methods. The statistical methods include descriptive analysis and description of the original data, analysis of the relationships (correlation and regression analysis, factor analysis, dispersion analysis), multivariate statistical analysis (clustering analysis, component analysis, discriminant analysis, multivariate regression analysis, canonical correlation etc.), the analysis of time rows (dynamic modeling and forecasting). The group of cybernetic methods include artificial neural networks (pattern recognition, clustering, forecasting), evolutionary programming (including algorithms of the group account of arguments), genetic algorithms (optimization), associative memory (search of analogues, prototypes), fuzzy logic, decision trees, systems for handling expert knowledge [3, 4]. In modern recommendation systems approaches using data mining algorithms are frequently applied to form predicative models. In particular, cluster analysis is widely used in this area.

In order to identify user preferences recommender systems collect and analyze large amount of data about users, including their geographic location, time spent on the various pages of the target resource, links clicked, items purchased and much more. The collected user activity data is characterized by large volume and diversity, as well as rapid change (update). Traditional databases are hardly applicable for working with this data because of its large amount and high performance requirements [5]. Usually the so-called NoSQL data management system (HBase, Cassandra) are used. Their characteristic feature is the rejection of the transactions, almost linear scalability, high-speed query processing, and absence of a rigid schema of data.

1.3 Methods and Problems of Recommender Systems

Adaptation for a particular user is a very difficult task as it is necessary to take into account user's personal uncertainty and spontaneity within a specific Internet resource, as well as many uncertainties connected with the peculiarities of the functioning of the Internet. In the context of content personalization (as well as prediction, identification of similar preferences and resource groups) the task is to process the collected data to identify specific patterns enabling to draw conclusions about the specific users' preferences. Thus, the main purpose of processing the data about user activity is to extract information that can be used for solving the following tasks:

- clustering of information units (in the case of e-commerce - goods);
- creating behavioral profiles of users (a generalization of the entire set of explicit and implicit data);
- clustering of user profiles;
- creating a personalized information offer for a user from a set of information units.

Thus, the main task of recommender system is creating a content maximally meeting the expectations (including implicit) of specific users. In order to solve this problem in most of modern recommendation systems the following basic approaches are used: collaborative filtering (CF) and content filtering (content-based filtering, CbF) [6]. Content filtering method focuses on identifying items similar to those that user was already interested in. This method considers the model of user behavior and characteristics (content) of the items the user was interested in. Items with similar characteristics (content) are identified to make recommendations. To be effective, the method of content filtering usually requires a detailed description of the characteristics of objects (e.g. in Music Genome Project music analyst evaluates each composition basing on hundreds of different musical characteristics), as well as information about a particular user (e.g. answers to specific questions in the questionnaire).

The collaborative filtering method is based on the assumption about conservatism of user preferences (i.e. users similarly assessing certain objects are likely to evaluate in the same way new objects with similar characteristics) [7]. The recommendations are based on an automated multi-user collaboration and selecting (by filtration method) users demonstrating similar preferences or behavior patterns. Thus, the collaborative filtering method generates recommendations based on previous user behavior patterns and taking into account the behavior of users with similar characteristics.

The most popular e-commerce recommender systems use the following variants of collaborative filtering method, as well as their hybrids:

- collaborative filtering by analyzing the preferences of user groups with similar interests (User-User Collaborative Filtering, User-User CF);
- collaborative filtering by analyzing the relationships between objects (Item-Item Collaborative Filtering, Item-Item CF).

The main problems associated with the implementation and practical use of collaborative filtering algorithms are data sparseness, problem of "cold start", and scalability. In addition to the above-mentioned challenges there is a problem of the limitation of offer diversity. Recommendation systems using collaborative filtering tend to offer already popular products, which creates problems for the promotion of new goods and services [8].

User-User Method

In the method of User-User CF similarities between users are determined and n most frequently purchased goods by k most similar customers are recommended to the user. To evaluate the degree of similarity in terms of user preferences a variety of functions of similarity (metrics) can be used. The most popular among them are the following: Euclidean distance, cosine measure, Hamming distance, Pearson correlation coefficient, Tanimoto coefficient, Manhattan distance, and so on [5, 7]. Defining recommendations by User-User CF method implies the construction of a matrix of users' activity, each line describing the actions of a specific user related to particular object (category, product, service) on website. User actions can be indicated in various ways. For example, it may be binary information about visiting or not visiting a given resource by a given user, data rate (or number) of usage resource R by user U, value or rating that user U assigns to resource R, etc. Thus, each line of activity of the matrix is a rating vector corresponding to different categories of products (thematic user profile). User profile characterizes the degree of his interest in each product group. For each pair of "user – object (product, service, action)" in the matrix of activity proximity the value of the selected metrics is calculated [9].

In order to find recommendations for a particular user based on his profile three main approaches are used: based on proximity (memory based), based on model (model based) and hybrid approach (hybrid). In modern commercial systems hybrid approach and model based approach (clustering algorithms, Bayesian belief networks, latent semantic models) are most popular. Different clustering algorithms are often used to identify groups of objects (users, products, target groups) with similar properties. In particular, in [10] it is shown that the problem of identifying user groups is naturally related to clustering methods in use. Clustering of data can also be used to generate users profiles based on the information about the actions of each user, and then to form groups of users based on their profiles.

Item-Item Method

Item-Item CF historically emerged as an alternative to User-User CF techniques designed to improve the performance of recommender systems for the stores, where the number of customers significantly exceeds the number of products in the catalog [11]. Initially, this method was proposed by Amazon to address the following key issues of User-User CF approach: the problem of cold start and the problem of frequent updates of user activity data. The problem of cold start significantly reduces the quality of the recommendation system due to the lack of data regarding the preferences of new (or not enough active) users. The problem of frequent updates of user activity data (in the case of Amazon we are talking about millions of customers) dramatically affects the perform-ance of the advisory system as a whole.

The basic idea of the Item-Item CF is to group information items (goods, services, activities) having similar assessment by users (ratings). The recommendations are generated in line with the following principle: the user who praised the object X is offered object Y, which is appreciated by other users who also praised object X. The use of Item-Item CF can improve the quality of recommendations for new users (no critical dependence on user preferences data) and also considerably enhance performance of recommender system when the number of users significantly exceeds the number of objects (object characteristics change less frequently). Unlike User-User CF, where calculating the degree of proximity in the "user – object" pairs is usually performed in real time (as the current transaction data becomes available only at the time of making recommendations), for Item-Item CF the closeness of the analyzed product to all other products can be calculated in a pending scheduled mode since vector ratings of all prod-ucts are available before recommendations. Thus, due to the possibility of delayed data processing, Item-Item CF is more efficient in terms of time of generating recommen-dations, the quality of advice on average is not worse than that based on user profiles analysis. The same metrics as in case of the pairs of "user – object" may be applied to calculate pairwise affinity of information units (a modified cosine or cosine measures are often used). To find recommendations based on a matrix of objects weighting functions and the regression-analysis methods are commonly used. One of the promising methods for solving the Item-Item CF problem is Item2Vec method [12].

Problems and Solutions for Recommender Systems Methods

The main drawback of applying Item-Item CF is the problem of creating of a reliable ranking of information units (products). Such a rating may be formed with the help of a large number of users who either regularly buy different goods (which directly affects the rating), or are explicitly involved in rankings (e.g. by evaluating goods or filling in the questionnaire). For smaller internet shops with a limited number of visitors, this approach can hardly be applied. At the same time User-User CF in terms of quality of recommendations highly depends on the availability of data about user activity and the problem of cold start is important for it. In general, it turns out that separately Item-Item CF and User-User CF use only a portion of the information collected by recommender systems about user activity and preferences when making recommendations. It may be more efficient to use hybrid approaches. For example, in [13] an algorithm for combined filtering is proposed, the main idea being obtaining a valuation of the unknown rating as the weighted sum of the estimates based on filtering for transactions, filtering for goods and the mixed filtration (on the basis of ratings of similar goods in similar transactions).

The structure of the paper is the following: in Sect. 2 hypothesis for investigation are formulated and also object of experiments is described; in Sect. 3 approaches to clustering of information items are presented; in Sect. 4 proposed methods of clustering user profiles are discussed; in Sect. 5 the evaluations of the results of experiments are presented and some conclusions are done. This article was prepared on the basis of reports [14, 16], however the content was significantly revised and extended. The first half of Sect. 1.1, almost the entire Sect. 1.2, subsection "Problems and solutions for recommender systems methods" of Sect. 1.3 are new. The text of Sect. 2 is also new, except for the description of the test data set. The analysis of the results in Sect. 3 was significantly extended. Section 5 describing data from practical experiments is completely new.

2 Hypothesis and a Base for Research

The authors of this paper propose to combine both of the above considered methods of collaborative filtering in order to better use the available information about the users and information units, and first of all implicit user data.

The essence of the hybrid approach is in simultaneous usage of ratings obtained via both Item-Item CF and User-User CF when developing recommendations. For new users or users with a non-representative history of visits we propose to generate recommendations basing on data about the information items similarity (Item-Item CF). Thus the problem of cold start is solved, and the quality of recommendations for inactive users and users with non-defined preferences is improved. With the accumulation of data on user preferences and generation of his behavioral profile it is proposed to give priority to estimates derived using User-User CF basing on the hypothesis that given the availability of qualitative user behavioral profiles, this method can predict their preferences more accurately. In this case recommendations obtained using User-User CF may be extended in case of necessity by offers of information units based on their popularity ratings.

Formulated hypothesis and working capacity of methods have been experimentally investigated in Thaisoap online store. The shop is focused on selling natural Thai cosmetics and coconut oil. Catalog of the store goods contains more than 1500 items segmented into180 classes (44 root classes, 136 subclasses). Every day, as per information for 2015, the store is visited by an average of about 1 500 visitors and each of them spends there 11 min on average (each visitor clicking 28 links on average). For theoretical research we used test-sample data for the fourth quarter of 2015, pilot studies were conducted in February – May of 2016. In both cases, the catalog of goods was unchanged. In addition, we used statistical data of the analytical system Yandex.Metrika (http://metrika.yandex.ru).

3   Clustering Information Items (Goods)

In order to address the problem of generating recommendations with relevant information given the lack of knowledge about user preferences the authors proposed to use the method based on the calculation of proximity of pairs and subsequent grouping (clustering) information units based on data about users who view multiple products in a row [14]. In the absence of necessary data conventional classifiers are recommended to be used, they take into account the prices and parameters of the objects, the list of "What's new," and the matrix "Those who buy this product also bought" (accessories that complement the basic purchase). In this approach, an explicit participation of Internet shop users in creating good rating is not required.

The first step of the algorithm is to construct a similarity matrix of information units, where all information units of online store are put in columns and lines. The matrix is filled according to the following rule: if the user successively viewed two products, the similarity weight in the matrix for these two items is incremented by 1. For processing the matrix in order to identify groups of information units that are close in their similarity scores, the modern productive Affinity Propagation algorithm was selected from all known clustering algorithms as a result of simulation conducted. One of the advantages of this algorithm is no need for a preliminary assessment of the optimal number of clusters [15].

By using statistical package R the treatment of test data array of online store Thaisoap was conducted: similarity matrix was built around all the available time (see Fig. 1) and its processing by the algorithm Affinity Propagation was carried out.

	p43	p44	p45	p46	p47	p48	p49	p50	p51	p52	p53	p54
p43	1.00000000	0.04347826	0.00000000	0.04347826	0.00000000	0.04347826	0.04347826	0.04347826	0.30434783	0.04347826	0.08695652	0.00000000
p44	0.04347826	1.00000000	0.00000000	0.04347826	0.00000000	0.00000000	0.08695652	0.04347826	0.04347826	0.08695652	0.13043478	0.04347826
p45	0.00000000	0.00000000	1.00000000	0.00000000	0.00000000	0.00000000	0.00000000	0.00000000	0.00000000	0.00000000	0.00000000	0.00000000
p46	0.04347826	0.04347826	0.00000000	1.00000000	0.00000000	0.00000000	0.04347826	0.04347826	0.04347826	0.04347826	0.04347826	0.00000000
p47	0.00000000	0.00000000	0.00000000	0.00000000	1.00000000	0.00000000	0.00000000	0.00000000	0.00000000	0.00000000	0.00000000	0.00000000
p48	0.04347826	0.00000000	0.00000000	0.00000000	0.00000000	1.00000000	0.13043478	0.00000000	0.00000000	0.00000000	0.08695652	0.00000000
p49	0.04347826	0.08695652	0.00000000	0.04347826	0.00000000	0.13043478	1.00000000	0.04347826	0.08695652	0.04347826	0.08695652	0.00000000
p50	0.04347826	0.04347826	0.00000000	0.04347826	0.00000000	0.00000000	0.04347826	1.00000000	0.04347826	0.04347826	0.04347826	0.00000000
p51	0.30434783	0.04347826	0.00000000	0.04347826	0.00000000	0.00000000	0.08695652	0.04347826	1.00000000	0.17391304	0.13043478	0.00000000
p52	0.04347826	0.08695652	0.00000000	0.04347826	0.00000000	0.00000000	0.04347826	0.04347826	0.17391304	1.00000000	0.30434783	0.00000000
p53	0.08695652	0.13043478	0.00000000	0.04347826	0.00000000	0.08695652	0.08695652	0.04347826	0.13043478	0.30434783	1.00000000	0.00000000
p54	0.00000000	0.04347826	0.00000000	0.00000000	0.00000000	0.00000000	0.00000000	0.00000000	0.00000000	0.00000000	0.00000000	1.00000000
p55	0.00000000	0.00000000	0.00000000	0.04347826	0.04347826	0.00000000	0.00000000	0.00000000	0.00000000	0.00000000	0.00000000	0.00000000
p56	0.00000000	0.00000000	0.00000000	0.00000000	0.00000000	0.00000000	0.00000000	0.00000000	0.00000000	0.00000000	0.00000000	0.00000000
p58	0.17391304	0.00000000	0.00000000	0.00000000	0.00000000	0.00000000	0.08695652	0.00000000	0.00000000	0.00000000	0.08695652	0.00000000
p60	0.04347826	0.08695652	0.04347826	0.08695652	0.00000000	0.00000000	0.04347826	0.04347826	0.04347826	0.04347826	0.08695652	0.00000000
p61	0.00000000	0.00000000	0.00000000	0.00000000	0.00000000	0.00000000	0.00000000	0.00000000	0.00000000	0.00000000	0.00000000	0.00000000
p64	0.00000000	0.00000000	0.00000000	0.00000000	0.00000000	0.00000000	0.00000000	0.00000000	0.00000000	0.00000000	0.00000000	0.00000000
p65	0.08695652	0.04347826	0.00000000	0.04347826	0.00000000	0.04347826	0.04347826	0.04347826	0.08695652	0.08695652	0.08695652	0.00000000
p66	0.04347826	0.04347826	0.00000000	0.04347826	0.00000000	0.00000000	0.00000000	0.08695652	0.04347826	0.08695652	0.04347826	0.00000000
p67	0.04347826	0.04347826	0.00000000	0.04347826	0.00000000	0.00000000	0.04347826	0.04347826	0.08695652	0.04347826	0.04347826	0.00000000
p71	0.00000000	0.00000000	0.00000000	0.00000000	0.00000000	0.00000000	0.00000000	0.00000000	0.00000000	0.00000000	0.00000000	0.00000000
p72	0.00000000	0.00000000	0.00000000	0.00000000	0.00000000	0.00000000	0.00000000	0.00000000	0.00000000	0.00000000	0.00000000	0.00000000
p73	0.00000000	0.00000000	0.00000000	0.00000000	0.00000000	0.00000000	0.00000000	0.00000000	0.00000000	0.00000000	0.00000000	0.00000000
p75	0.08695652	0.04347826	0.00000000	0.04347826	0.00000000	0.00000000	0.04347826	0.04347826	0.13043478	0.08695652	0.08695652	0.00000000

Fig. 1. Similarity of goods matrix

The results of the algorithm are shown in Fig. 2 in the form of a clustered heat map (size of map is 1522×1522). The prevailing of one color on the map is due to the fact that in a test data sample for most pairs of goods the similarity score is not picked (i.e. users were not interested in these products during the period examined in the test sample).

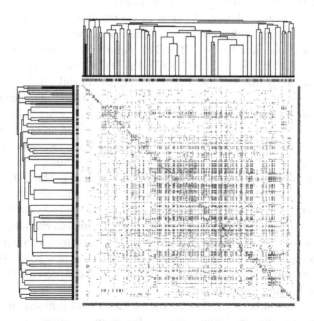

Fig. 2. Cluster heat map (Color figure online)

The algorithm identified 64 clusters, the largest being clusters number 5 (75 objects), 8 (44 objects), 10 (30 objects), 19 (27 objects), and 55 (31 objects). The quality of the algorithm can be estimated by the example of cluster number 5, its description is given in Table 1.

Table 1. Detailing of cluster number 5.

Cluster	Reference information item	Examples of products from the cluster
ID: 5 Size: 75	ID: 76 Unrefined 100% coconut massage oil "Citronella" Tropicana, 100 ml	ID: 43 Coconut oil Tropicana 1 L, unrefined ID: 51 Coconut oil unrefined Tropicana in apothecary bottle, 90 ml ID: 466 Revitalizing Coconut Body Lotion Tropicana "Sweet Coconut" (without parabens), 200 ml ID: 624 Mask exfoliant for the face "Marine collagen"Artiscent, 100 ml ID: 1234 Mini set of shampoo and hair conditioner "Golden silk with mulberry extract"

In particular, it is clear that for the reference information item (massaging coconut oil) the products based on coconut oil or indirectly associated with creams and oils for body care have been included in the similarity cluster.

Consideration of other clusters obtained as the result of the algorithm application shows that goods having similar consumer characteristics were put in separate groups. For example, the cluster number 55 ("Blue Thai balm for varicose veins") was formed mainly with balms ("Traditional small RED Thai balm (for healing massage) Korn Herb", etc.) and nail polishes ("Antifungal nail polish "Demikten"" and etc.), having treatment character. In the cluster number 10 ("Set "Healthy Hair" Tropicana") products for hair and body were included ("Hair Serum "Romance"", "Set "Magnificent hair" Tropicana", etc.). A product that interested potential buyer in an online store catalog may be used as a reference information item. In this case the recommendations of the system will be based on products with the closest estimates of similarity to the reference item (i.e. belonging to the corresponding similarity cluster). Thus, if a new visitor chooses in the catalog massage coconut oil "Citronella Tropicana", the recommender system will prompt him to view coconut oil Tropicana, coconut oil unrefined Tropicana in apothecary bottle, revitalizing coconut body lotion Tropicana etc. Depending on the configuration of an advisory service goods offering (based on the method Item-Item CF) may be supplemented by currently the most popular goods or new products that require promotion (for example, one of the five slots for recommended products can be assigned to promote new products from the massage oils line).

In general, it should be noted that the use of Item-Item CF based on evaluation of the information units similarity allows to provide recommendations intuitively clear for shop visitors, not having data on their preferences. For example, when choosing a balm for varicose veins the recommender system will propose to view also a traditional balm for healing massage and antifungal nail polish, which in general will fairly accurately correspond to the topic of search given by visitor (treatment cosmetics on the basis of coconut oil).

4   Clustering User Profiles

With the accumulation of user activity data it becomes possible to create behavioral profiles and use User-User CF to generate recommendations more relevant to the specific users' preferences. As a result of the performed simulation, it was found that K-means clustering [16, 17], which takes into account current limits on speed and data volume, is the most suitable method to identify groups of users with similar preferences (User-User CF). In general, this method is simple to implement and is scalable. Computational complexity of the algorithm is $O(nkl)$, where n is the number of objects, k is the number of clusters, l is the number of iterations. One of the drawbacks of this method is the need to pre-set the number of clusters to split. Nevertheless, it is possible to get rather simply the optimal number of clusters to split by minimizing the sum of intra-cluster distances. Another drawback of the K-means algorithm is direct dependence of its performance on the number of iterations. The number of iterations can be reduced by assigning near-optimal initial values to cluster centroids (this approach is used in the modified method of K-means - K-means++).

The first step of the algorithm is to construct a matrix of users' activity. In the test data set only the number of addressing particular categories of goods was available. Thus, each line of the activity matrix is a vector of estimates corresponding to different categories of products (thematic user profile). User profile characterizes the degree of his interest to each product group.

The next step (before identifying groups of users with similar characteristics via clustering techniques) is processing of the developed activity matrix in order to calculate pairwise distances between its elements. The Euclidean distance (geometric distance in the multidimensional space), which is one of the simplest to implement, was used as the distance metrics (similarity function). Figure 3 shows a histogram of distances obtained by processing the matrix with pairwise distances between objects using the statistical package R. Preliminary review of the results leads to the conclusion that preferences (vector activity) of online store Thaisoap users do not differ significantly (i.e., most likely that majority of users are interested in similar categories of goods).

One of the initial parameters for the K-means application is the number of clusters to split the analyzed dataset. To determine the optimal number of clusters a special study was carried out. As a criterion the intercluster sum of the squares of the distances that needed to be minimized was used. Figure 4 shows the results of calculation of the inter-cluster sum of squared distances on the elbow method (Elbow method). The study of the test data set has determined that the optimal number of clusters for the given test array is equal to 30.

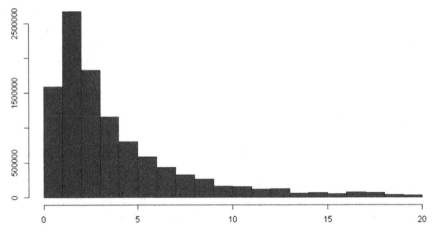

Fig. 3. Histogram of distances

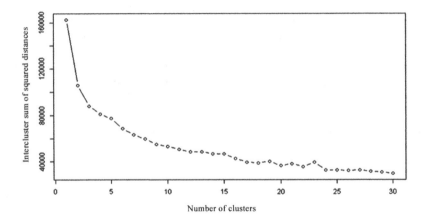

Fig. 4. Analysis of the number of clusters

Thus, all necessary parameters for clustering data on user activity matrix were obtained. Figure 5 graphically shows the result of the clustering algorithm work.

The result of the clustering algorithm allowed figuring out several largest clusters of users with the numbers 1, 9, 20, 21 and 22. Users from the cluster number 1 (cluster size 1384) show weak preference for all categories of goods. It is possible that the cluster number 1 includes visitors who came without a specific purpose, e.g., just to get acquainted with the assortment of goods offered. Users belonging to the cluster number 9 (size 180) show clear preference for cat9 category ("Face"). Members of the cluster number 20 (size 1366) prefer cat7 category ("Coconut Oil"). The sum of squared distances in relation to other clusters is small, indicating the small difference between the objects within the cluster. Members of the cluster number 21 (size 622) are also demonstrating preference for cat7 category ("Coconut Oil"). Members of the cluster number 22 (size 180) prefer categories cat47 ("MUST HAVE Winter 2016") and cat49

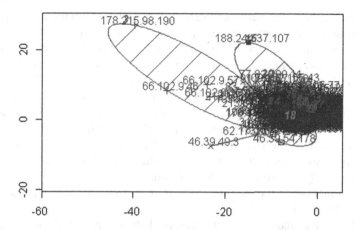

Fig. 5. The result of the algorithm of user profiles clustering

("HITS of our store"). Carrying out calculations based on the data from other weeks gave similar results.

Thus, clusters obtained as the result of processing the original user activity data can be used by recommender system for generating offers. A quality of recommendations will depend on the adequacy of the user preference data (i.e., on the quality of the generated behavioral profiles). Information units having the smallest distance metrics in relation to the reference product will be recommended to users (reference product being the one that interested visitor at a particular time). Within the hybrid approach the recommendations may be extended in case of necessity by suggestions of information items having close similarity (Item-Item CF). The need to extend recommendations received via the method User-User CF is most likely to arise in relation to users who do not have clearly expressed preferences.

5 Evaluation of Clustering Results

Due to the lack of opportunities to assess the hybrid approach in the long run, it was decided to conduct split-testing, i.e. to organize comparison of Thaisoap online store users behavior based on four scenarios: with no personalized information recommendation, with only recommendations based on Item-Item CF, with only User-User CF recommendations, and personalized recommendations based on the hybrid approach developed by the authors.

Studies have been conducted on the basis of an experimental software package that implemented each of the four scenarios. Data on one month period from February to May 2016 was taken for testing each scenario. At that time the catalog of goods remained unchanged (i.e. the product line recommended to store visitors was the same). One of the conditions of the comparative analysis was the work with "a clean slate", i.e. at the beginning of each month of the period the data about user preferences or ratings of information items accumulated over previous periods was not used.

In the first scenario 1860 visitors were analyzed in total, 140 of them made purchases (7.5% conversion). Average number of items in the basket was 1.5. When navigating through the site visitors made on average 36 clicks in one session and spent on site about 16 min. In the second scenario there were 1908 visitors, 164 of them making purchases (8.6% conversion). Average number of items in the basket was 2.1. When navigating through the site visitors made on average 31 clicks in one session and spent on the site about 12 min. In the third scenario 1873 visitors were analyzed, 145 people making purchases (7.7% conversion). Average number of items in the basket was 1.9. When navigating through the site visitors made 32 clicks in one session and spent on the site about 14 min on average. In the latter scenario there were 2011 visitors, 183 of them made purchases (9.1% conversion). Average number of items in the basket was 2.4. When navigating through the site visitors made on average 28 clicks in one session and spent on the site about 11 min.

Thus, the recommender system based on the hybrid approach allowed to reach the greatest increase of the indicators under review. For example, the conversion rate in comparison with the first scenario increased by more than 1.5%, and the average number of items in the basket increased from 1.5 to 2.4 (i.e. more than on 60%).

In general, it should be noted that the effectiveness of recommender system strongly depends on the type of goods sold, on the features of the algorithms used and on the quality of recommender system implementation. In certain situations recommender system does not provide a significant increase in conversion rates and average number of items in the basket. Nevertheless, it remains useful as it offers a convenient and user-friendly method for the search of the goods consumers are interested in. The system creates a comfortable environment and strengthens the loyalty of visitors to the online resource, which already in the medium term can lead to increased returning visits and creation of pool of loyal customers that ensure stable revenue growth of the online store.

Acknowledgements. This work was supported by the Ministry of Education and Science of the Russian Federation, grant no. RFMEFI60414X0139.

References

1. Walker, L.: Amazon Gets Personal With E-Commerce. Washington Post Staff Writer Sunday, Page H1, 8 November 1998. http://www.washingtonpost.com/wp-srv/washtech/daily/nov98/amazon110898.htm
2. Philippov, S., Zakharov, V., Stupnikov, S., Kovalev, D.: Approaches to improve the pertinence of information in the media services on the basis of big data processing. In: Selected Papers of the XVII International Conference on Data Analytics and Management in Data Intensive Domains (DAMDID/RCDL 2015). CEUR Workshop Proceedings, vol. 1536, Obninsk, Russia, pp. 114–118, 13–16 October 2015
3. Parsaye, K.: OLAP and data mining: bridging the gap. Database Program. Des. **10**(2), 30–37 (1997)
4. Parsaye, K.: Characterization of data mining technologies and processes. J. Data Warehouse. **1**, 13–18 (1998)

5. Philippov, S., Zakharov, V., Stupnikov, S., Kovalev, D.: Organization of big data in the global e-commerce platforms. In: Selected Papers of the XVII International Conference on Data Analytics and Management in Data Intensive Domains (DAMDID/RCDL 2015). CEUR Workshop Proceedings, vol. 1536, Obninsk, Russia, pp. 119–124, 13–16 October 2015

6. Jones, M.: Recommender systems. Part 1: Introduction to approaches and algorithms (2013). http://www.ibm.com/developerworks/library/os-recommender1/

7. Su, X., Khoshgoftaar, T.M.: A survey of collaborative filtering techniques. Adv. Artif. Intell. **2009**, 19 (2009). Article ID 421425

8. Fleder, D., Hosanagar, K.: Blockbuster culture's next rise or fall: the impact of recommender systems on sales diversity. Manag. Sci. **55**(5), 697–712 (2009)

9. Breykin, E.A.: Rekomendatel'naya sistema na osnove kollaborativnoy fil'tratsii. Molodoy uchenyy **13**, 31–33 (2015)

10. Marmanis, Kh., Babenko, D.: Algoritmy intellektual'nogo Interneta. In: SPb.-M., Simvol (2011)

11. Linden, G., Smith, B., York, J.: Amazon.com recommendations: item-to-item collaborative filtering. Industry report. IEEE Internet Comput. **7**, 76–80 (2003)

12. Barkan, O., Koenigstein, N.: Item2Vec: Neural Item Embedding for Collaborative Filtering (2016). arXiv preprint: arXiv:1603.04259

13. Goncharov, M.: Sistemy vyrabotki rekomendatsiy (2010). http://www.businessdataanalytics.ru/RecommendationSystems.htm

14. Philippov, S., Zakharov, V., Stupnikov, S., Kovalev, D.: Determination of similarity of information entities based on implicit user preferences in life-support recommender systems. In: Selected Papers of the XVIII International Conference on Data Analytics and Management in Data Intensive Domains (DAMDID/RCDL 2016). CEUR Workshop Proceedings, vol. 1752, Ershovo, Moscow Region, Russia, pp. 104–109, 11–14 October 2016

15. Frey, B.J., Dueck, D.: Clustering by passing messages between data points. Science **315**(5814), 972–976 (2007). doi:10.1126/science.1136800

16. Philippov, S., Zakharov, V., Stupnikov, S., Kovalev, D.: Clustering of user profiles based on real implicit data in e-commerce recommender systems. In: Selected Papers of the XVIII International Conference on Data Analytics and Management in Data Intensive Domains (DAMDID/RCDL 2016). CEUR Workshop Proceedings, vol. 1752, Ershovo, Moscow Region, Russia, pp. 98–103, 11–14 October 2016

17. Coates, A., Ng, A.Y.: Learning Feature Representations with K-means. Stanford University (2012). http://www.cs.stanford.edu/~acoates/papers/coatesng_nntot2012.pdf

On Data Persistence Models for Mobile Crowdsensing Applications

Dmitry Namiot[1(✉)] and Manfred Sneps-Sneppe[2]

[1] Faculty of Computational Mathematics and Cybernetics,
Lomonosov Moscow State University, Moscow, Russia
dnamiot@gmail.com
[2] Ventspils International Radioastronomy Centre, Ventspils University College,
Ventspils, Latvia
manfreds.sneps@gmail.com

Abstract. In this paper, we discuss various models and solutions for saving data in crowdsensing applications. A mobile crowdsensing is a relatively new sensing paradigm based on the power of the crowd with the sensing capabilities of mobile devices, such as smartphones, wearable devices, cars with mobile equipment, etc. This conception (paradigm) becomes quite popular due to huge penetration of mobile devices equipped with multiple sensors. The conception enables to collect local information from individuals (they could be human persons or things) surrounding environment with the help of sensing features of the mobile devices. In our paper, we provide a review of the data persistence solutions (back-end systems, data stores, etc.) for mobile crowdsensing applications. The main goal of our research is to propose a software architecture for mobile crowdsensing in Smart City services. The deployment for such kind of applications in Russia has got some limitations due to legal restrictions also discussed in our paper.

Keywords: Crowdsensing · Mobile · Cloud · Context-aware

1 Introduction

This article presents an extended and redesigned version of our paper presented on DAMDID/RCDL 2016 conference [1].

As per classical definition, mobile crowdsensing is a sensing paradigm, which is based on the power of the crowd mobile users (more widely - mobile devices) with the sensing capabilities [2]. It is illustrated in Fig. 1.

So, the sensor-enabled devices are the key enablers for mobile crowdsensing. The background for this process is very obvious. There are pure economical reasons behind this process. We see the increasing popularity of smartphones, already equipped with multiple sensors. So, why do not use them for collecting the local timely-based knowledge from the surrounding environment? We do not need to place sensor hardware and we do not need to create special networks for collecting data. That is why, for example, crowdsensing is extremely useful for Smart Cities where budgets are usually limited.

© Springer International Publishing AG 2017
L. Kalinichenko et al. (Eds.): DAMDID/RCDL 2016, CCIS 706, pp. 192–204, 2017.
DOI: 10.1007/978-3-319-57135-5_14

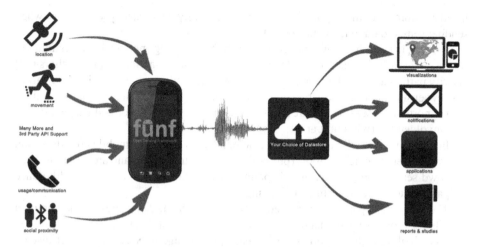

Fig. 1. Mobile crowd sensing [3]

There are several important issues that should be mentioned. Of course, in crowd-sensing, we can collect various data: location data, camera information, air pollution data, etc. In other words, it is everything that could be measured with a mobile device. And it is not only that could be done through the sensing features of the mobile device. For example, each smartphone has wireless network modules (Wi-Fi, Bluetooth, Bluetooth Low Energy). We can use information about "visible" (available) wireless networks (nodes, signal strength) as sensing data too. It lets create radio-maps for cities (buildings) and it could be used in physical web models [4]. In other words, the word "sensor" here is the synonym for the word "measurement".

Originally, mobile crowdsensing was about collecting data via smartphones only (as it is presented in Fig. 1). But now we can talk about wearable devices too, about cars for data collecting, etc. For example, our paper [5] presents network proximity models for cars. Classically, GPS data collected by cars have been used to analyze such problems as traffic congestion and urban mobility [6]. As per the latest vision for Internet of Things, we can collect data even from the individual itself – there are so-called cyber-physical systems [7]. Telecom operators, for example, are collecting cellular location information for every SIM-card, so, the mobile phone itself is a sensor collected data about owner's mobility [8].

There are two groups if crowdsensing challenges, usually mentioned in the scientific papers. They are user's anonymity and data sensing quality. The first problem is based on the fact that humans (mobile phone/wearables owners) participate in the process directly or indirectly. In this case, the performance and usefulness of crowdsensing sensor networks depend on the crowd willingness to participate in the data collection process. But we have noted above that the data can be collected not only from the owners of mobile phones. Also, as it was in our case, the information sources may be corporate mobile devices and municipal transport, so that there is no question of voluntariness.

A popular approach for preserving user privacy is depersonalization. It could be done via removing any user identifying attributes from the sensing data before sending it to the data store. Another approach is to use randomly generated pseudonyms when sending sensed data to the data store [9, 10].

As per [11], we have two categories for mobile sensing: personal and community sensing. Our tasks (Smart City) belong to community sensing. Some of the authors [12] classify participatory and opportunistic sensing. It depends on user participation. Participatory sensing includes the active involvement of users (participants) to contribute data. And opportunistic sensing is more autonomous and should be performed with a minimal user involvement (without it at all).

In our paper we will target another challenge, data stores for mobile crowdsensing. We will present a review of tools (preferably Open Source tools) and architectures used in crowd sensing projects. From this point of view (data stores) there is no difference for participatory and opportunistic sensing.

The rest of our paper is organized as follows. In Sect. 2, we present the common models for crowd sensing data architectures, discuss local databases and cloud-based solution, highlight the importance of stream processing. In Sect. 3, we will discuss crowdsensing for multimedia data (for video applications). In Sect. 4, we discuss back-ends for Smart Cities. In Sect. 5, we discuss local deployments. Our review has been produced as a part of a research project on Smart Cities and applications in Lomonosov Moscow State University. The main goal of this review is to propose the software architecture for mobile crowdsensing in Smart City environment. We note also that the software architecture for the proposed system must satisfy the existing law restrictions in the Russian Federation, which will be discussed below.

2   Basic Architecture for Mobile Crowdsensing

What are the typical requirements for mobile crowdsensing applications? A good summary can be found in [13]. The basic requirements are as follows:

- provide a minimal intrusion in client devices. The mobile device computing overhead always must be minimized. This requirement should cover all the stages: active state (passing data to data store) and passive state (waiting for new sensing data);
- provide fast feedback and minimal delay in producing stream information. Sometimes it depends on sensors (e.g., for asynchronous sensing [14]), but in general the latency should be minimized;
- provide open interfaces;
- provide security;
- support complete data management workflow, from data collection to communication;

Due to complexity of sensing collecting process, some models propose to use local databases for accumulating data on mobile devices and then replicate them for the processing. This schema is illustrated in Fig. 2 [13].

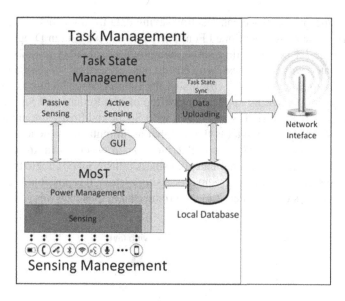

Fig. 2. Local database for sensing

This solution is very easy to implement and there are several options to choose. For example, Android platform offers several options for local data persistence:

- Shared Preferences. This option stores private primitive data in key-value pairs.
- Internal Storage. This option stores private data on the device memory.
- SQLite Databases. It stores structured data in a private database.

SQLite is the most often used solution here. It is a self-contained, embeddable, zero-configuration SQL database engine [15]. For example, Open Source Funf package from MIT [16] saves sensing info in SQLite database (Fig. 3).

Fig. 3. Funf datastore [17]

For crowdsensing system we will get a set of local databases. Accordingly, it will need some process of unification of local data and use a kind of Extract-Transform-Load (ETL) script [18]. Naturally, this model is not suitable for real time processing. As an intermediate improvement, we can mention saving data from sensors in a cloud-based data store. E.g., the above mentioned Funf package can save data in Dropbox.

The real-time (or near real-time) processing by the scalability reasons in the most cases is associated with some messaging bus. In this connection, we should mention so-called Lambda Architecture [19]. Originally, the Lambda Architecture is an approach to building stream processing applications on top of MapReduce and Storm or similar systems (Fig. 4). Nowadays it is associated with Spark and Spark streaming too [20]. The main idea behind this model is the fact that an immutable sequence of records is captured and fed simultaneously (in parallel) into a batch system and a stream processing system. So, developers should implement business transformation logic twice, once in the batch system and once in the stream processing system. It is possible to combine the results from both systems at query time to produce a complete answer [21].

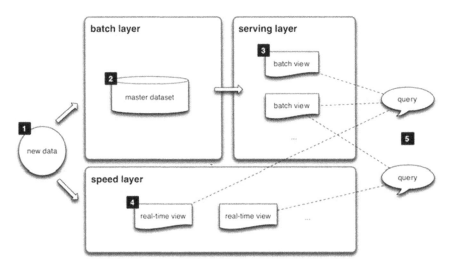

Fig. 4. Lambda architecture [21]

The Lambda Architecture targets applications built around complex asynchronous transformations that need to run with low latency. Any batch processing takes the time. In the meantime, data has been arriving and subsequent processes or services continue to work with old information. The Lambda Architecture offers a dedicated real-time layer. It solves the problem with old data processing by taking its own copy of the data, processing it quickly and storing it in a fast store. This store is more complex since it has to be constantly updated.

One of the obvious disadvantages is the need for duplicating business rules. Practically, the developers need to write the same code twice for real-time and batch layers. One proposed approach to fixing this is to have a language or a framework that abstracts over both the real-time and batch frameworks [22].

The database (data store) design for stream processing has its own specific [23]. In general, there are two options: storing every single event as it comes in (for sensing, it means storing every single measurement), or storing an aggregated summary of the measurements (events).

The big advantage of storing raw measurements data is the maximum flexibility for the analysis. Hadoop could be used for the full dump. However, the summary (preprocessed data) also has its use cases, especially when we need to make decisions or react to things in real time. Implementing some analytical methods with raw data storage would be incredibly inefficient because we would be continually re-scanning the history of measurements. So, the both options could be useful, they just have different use cases.

As the next step in Lambda architecture, we could mention so-called Kappa architecture [24], where everything is a stream. In our opinion, it is the most suitable approach for sensing data persistence (Fig. 5).

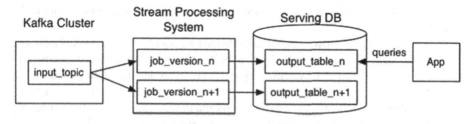

Fig. 5. Kappa architecture

There are several Open Source solutions for data streaming support, e.g., Apache Flink, Flume, Chukwa [19]. However, the most used system (at least, in sensing-related tasks) is Apache Kafka. In general, publish-subscribe architecture is the most suitable approach for scalable crowd sensing applications. Apache Kafka is a distributed publish-subscribe messaging system. It is designed to provide high throughput persistent scalable messaging with parallel data loads into Hadoop. Its features include the use of compression to optimize performance and mirroring to improve availability, scalability. Kafka is optimized for multiple-cluster scenarios [25]. Technically, there are at least three possible message delivery guarantees in publish-subscribe systems:

- At most once. It means that messages may be lost but are never redelivered;
- At least once. It means messages are never lost but may be redelivered;
- Exactly once. It means each message is delivered once and only once.

As per Kafka's semantics when publishing a message, developers have a notion of the message being "committed" to the log. As soon as a published message is committed, it will not be lost. Kafka is distributed system, so it is true as long as one broker that replicates the partition to which this message was written is still alive. In the same time, if a crowdsensing client (producer in terms of publish-subscribe systems) attempts to publish a new measurement and experiences a network error, it cannot be sure when this error happens. Did it happen before or after the message was committed? The most natural reaction for the client is to resubmit the message. It means that we could not guarantee the message had been published exactly once. To bypass this limitation we need some sort of primary keys for inserted data, which is not easy to achieve in distributed systems. For crowdsensing systems, we can use producer's address (e.g. MAC-address or IMEI of a mobile phone) as a primary key.

Kafka guarantees at-least-once delivery by default. It also allows the user to implement at most once delivery by disabling retries on the producer and committing its offset prior to processing a batch of messages. Exactly-once delivery requires co-operation with the destination storage system (it is some sort of two-phase commit).

In connection with Kafka, we would like to highlight two approaches. The rising popularity of Apache Spark creates the big set of projects for Kafka-Spark integration [26]. And the second approach is the relatively recently introduced Kafka Streams. Kafka models a stream as a log, that is, a never-ending sequence of key/value pairs. Kafka Streams is a library for building streaming applications, specifically applications that transform input Kafka topics into output Kafka topics (or calls to external services, or updates to databases, or whatever). It lets you do this with concise code in a way that is distributed and fault-tolerant [27].

On the client side for crowd sensing applications we could recommend the originally proposed by IBM Quarks System. Now it is Apache Edgent, previously known as Apache Quarks [28]. Quarks System is a programming model and runtime that can be embedded in gateways and devices. It is an open source solution for implementing and deploying edge analytics on varied data streams and devices. It can be used in conjunction with open source data and analytics solutions such as Apache Kafka, Spark, and Storm.

As for the future, there is an interesting approach from a new Industry Specification Group (ISG) within ETSI, which has been set up by Huawei, IBM, Intel, Nokia Networks, NTT DOCOMO and Vodafone. The purpose of the ISG is to create a standardized, open environment which would allow for efficient and seamless integration of applications from vendors, service providers, and third-parties across multi-vendor Mobile-edge Computing platforms [29]. This work aims to unite the telecom and IT-cloud worlds, providing IT and cloud-computing capabilities within the Radio Access Network. Mobile Edge Computing proposes co-locating computing and storage resources at base stations of cellular networks. It is seen as a promising technique to alleviate utilization of the mobile core and to reduce latency for mobile end users [30].

We think also that 5G networks should bring changes to the crowdsensing models. In is still not clear, what is a killer application for 5G. One from the constantly mentioned approaches is so-called ubiquitous things communicating. The hope is that 5G will provide super fast and reliable data transferring approach [31]. Potentially, it could change the sensing too. 5G should be fast enough, for example, to constantly save all sensing information from any mobile device in order to use them in ambient intelligence (AMI) applications [32]. Actually, in this model crowdsensing is no more than a particular use-case for ambient intelligence. But at the moment, these are only theoretical arguments.

3 Crowdsensing for Multimedia Data

In this section, we would like to discuss crowdsensing for video data. The practically considered task was the processing of data security cameras. The key question here is cloud storage for video data. Almost all existing projects related to media data use Amazon Simple Storage Service (S3) (Fig. 6).

Fig. 6. Amazon S3 storage [33]

The storage is based on buckets. Buckets and objects (e.g., videos) are resources managed via Amazon S3 API. The typical applications stores media objects in S3 buckets and keeps keys for them in a separate database (it could be a relational database or NoSQL key-value store). There are many open source applications based on Amazon S3 API, but the key question here is Amazon S3 or its analogs. In Russia, one can face law restrictions with public clouds. As per law, personal data should be stored locally. And since the word "personal" has a very broad interpretation, in practice it means that all locally obtained data should be stored locally too. From existing Amazon S3 analogs we know about partial implementation from Selectel [34]. So, in our opinion, it is a very prospect task for Russia to select some Open Source platform for IaaS and build an own cloud. There are several prototypes for the beginning. As Open Source platforms candidates in this area, we can mention, e.g., CloudStack [35]. Apache CloudStack is an open source cloud computing software, which is used to build Infrastructure as a Service (IaaS) clouds by pooling computing resources. Apache CloudStack manages computing and networking as well as storage resources.

Eucalyptus (Elastic Utility Computing Architecture for Linking Your Programs To Useful Systems) [36] is free and open-source computer software for building Amazon Web Services (AWS)-compatible private and hybrid cloud computing environments.

The OpenStack project [37] is a global collaboration of developers and cloud computing technologists producing the open standard cloud computing platform for both public and private clouds. In our opinion, it is a most suitable choice. OpenStack has a modular architecture with various code names for its components. The most interesting component (in the context of this paper) is OpenStack Compute (Nova). It is a cloud computing fabric controller, which is the main part of an IaaS system. It is designed to manage and automate pools of computer resources and can work with widely available virtualization technologies. It is an analog for Amazon EC2.

OpenStack Object Storage (Swift) is a scalable redundant storage system [38]. With Swift, objects and files are written to multiple disk drives spread throughout servers in the data center, with the OpenStack software responsible for ensuring data replication and integrity across the cluster. It lets scale storage clusters scale horizontally simply by adding new servers. Swift is responsible for replication its content.

In our opinion, the cloud-based solution for video data in Smart City applications is a mandatory part of eco-system and OpenStack Swift is the best candidate for the platform development tool.

The importance of cloud-based video services is confirmed by the industry movements. For example, we can mention here IBM's newest (2016) Cloud Video Unit business [39]. As a good example (or even a prototype for the development), we can mention also Smartvue applications [40]. In our opinion, the video processing for data from moving cameras (e.g., surveillance cameras in cars) is a new hot crowdsensing area in Smart Cities.

4   Smart Cities Back-Ends

There are several big Open Source projects directly target back-ends for Smart Cities applications. Firstly, it is FIWARE Cloud [41] (Fig. 7).

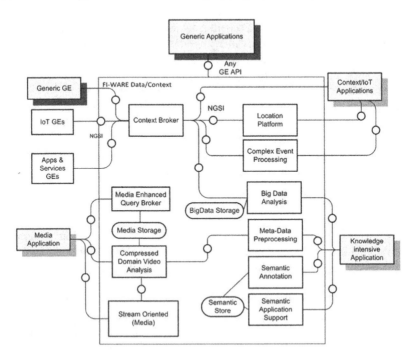

Fig. 7. FIWARE Cloud [41]

Technically, FIWARE supports third-party components (enablers). One of them, for example, proposes generic stream processing. This enabler [42] proposes public API for creating person-to-person services (e.g. video conferencing, etc.), person-to-machine services (e.g. video recording, video on demand, etc.) and machine-to-machine services (e.g. computerized video-surveillance, video-sensors, etc.). However in terms of data storage, it relies on public clouds, like Microsoft Azure.

There is a specific problem with public clouds in Russia: there are almost no developers familiar with the subject. All Smart Cities related projects we know about in Russia always introduce own back-end system. And we do not know even about attempts to create (introduce) a common standard in this area.

In general, it is a perfect example of so-called MBaaS (Mobile Backend As A Service) model for providing the web and mobile app developers with a way to link their applications to backend cloud storage [43]. MBaaS provides application public interfaces (APIs) and custom software development kits (SDKs) for mobile developers. Also, MBaaS provides such features as user management, push notifications, and integration with social networking services. The key moment here is the simplicity for mobile developers. As soon as many (most) of crowdsensing applications rely on mobile phones, this direction is very promising for crowdsensing. Actually, the additional (to data storage) services are the key idea behind MBaaS.

As an Open Source product in this area, we can mention Apache Usergrid [44]. It is composed of an integrated distributed NoSQL database, application layer and client tier with SDKs for developers. Usergrid lets developers rapidly build the web and mobile applications. It provides basic services, such as user registration and management, data storage, file storage, queues, as well as retrieval features such as full text search and geolocation search to power common features for applications.

5   On Practical Use-Cases and Deployment in Russia

In this section, we will present two use cases for back-end systems in prototyping projects with one mobile telecom operator. Firstly, it is wireless proximity information collection. Source data are network fingerprints (list of wireless nodes with signal strength). Each fingerprint has got a time stamp and could be associated (in the most cases) with some geo-coordinates. In our prototype, we use the following chain: Kafka -> Spark Streaming -> Cassandra. Cassandra has been selected as a database suitable for time series. Most of the measurements (including network proximity too) are de-facto time series data (multivariate time series). With the above mentioned chain, we can ensure the compliance with all applicable local restrictions: the personal data will be stored on the territory of the Russian Federation (all the above-mentioned components could be placed in local data centers) and Open Source components provide the absence of claims from the import-substitution point of view (this schema does not use any imported commercial software). Such a bundle is in line with modern approaches, so we can update components, reuse existing open source solutions for them, and participate in developers activities (in Open Source communities around the above-mentioned components).

The second example is much less successful. The idea of the application is data accumulation for dash cameras in vehicles. Data-saving entities are geo-coded media data (so-called ONVIF stream). So, it is crowdsensing for media data. The business value here is absolutely obvious. The existing city cameras are static and they cover predefined areas only, whereas users (cars) in the city can cover all the areas dynamically. De-facto standard for media data in a cloud is Amazon S3. But due to existing regulations (so-called personal data) it could not be used in Russia, because physically data will be saved outside of the country. We think that the "standard" solution is preferable, because there are many available components and systems based on Amazon S3 API. So, even the simulation of this API on the own data model lets reuse many existing software components. In our opinion, with the declared import-substitution and data localization regulations Amazon S3 analogue in Russia should be developed on the national level. Definitely, the only data centers building is not enough and we should talk about software tools for them too. At this moment, this topic is not discussed in Russia, which is a bit strange in our opinion. As a base for S3 analogue development, we can probably use OpenStack (OpenStack Swift). For example, the model from Rackspace Cloud Files (it is an analogue of Amazon S3 and based on OpenStack Swift) [45] could be reproduced.

Currently, Russia starts processes on the standardization of Internet of Things and Smart Cities. Of course, data persistence is an important part of such processes across the world and Russia could not be an exception here. Again, it looks reasonable to reuse already existing developments here. For example, we can mention such projects as oneM2 M or FIWARE (there is a review for domestic standards in our paper [46]). However standards in IoT (M2 M) do not provide dedicated data persistence solutions. They also rely on the existing cloud solutions. So, all the above-mentioned data saving regulations and restrictions are applicable here.

The next important trend is the strategy of vendors of sensors and other measuring devices. Many of them now include data storage as a part of "sensor". For example, Bluetooth tags Eddystone from Google include Google data storage too [47] (unlike iBeacons tags from Apple, for example). In our opinion, this trend will only rise, because data capturing lets vendors provide additional services. It means that with the existing restrictions for data locations, several classes of sensors will hardly be deployed in Russia.

References

1. Namiot, D., Sneps-Sneppe, M.: On crowd sensing back-end. In: DAMDID/RCDL 2016 Selected Papers of the XVIII International Conference on Data Analytics and Management in Data Intensive Domains (DAMDID/RCDL 2016), CEUR Workshop Proceedings, vol. 1752, pp. 168–175 (2016)
2. Tanas, C., Herrera-Joancomartí, J.: Users as smart sensors: a mobile platform for sensing public transport incidents. In: Nin, J., Villatoro, D. (eds.) CitiSens 2012. LNCS (LNAI), vol. 7685, pp. 81–93. Springer, Heidelberg (2013). doi:10.1007/978-3-642-36074-9_8

3. Internet of Things, Web of Data & Citizen Participation as Enablers of Smart Cities. http://www.slideshare.net/dipina/internet-of-things-web-of-data-citizen-participation-as-enablers-of-smart-cities. Accessed Jan 2017
4. Namiot, D., Sneps-Sneppe, M.: The physical web in smart cities. In: Advances in Wireless and Optical Communications (RTUWO 2015). IEEE Press, New York (2015)
5. Namiot, D., Sneps-Sneppe, M.: CAT - cars as tags. In: 2014 Proceedings of the 7th International Workshop on Communication Technologies for Vehicles (Nets4Cars-Fall). IEEE Press, New York (2014)
6. Massaro, E., et al.: The car as an ambient sensing platform. Proc. IEEE 105(1), 3–7 (2017)
7. Hu, X., Chu, T., Chan, H., Leung, V.: Vita: a crowdsensing-oriented mobile cyber-physical system. IEEE Trans. Emerg. Top. Comput. 1(1), 148–165 (2013)
8. Calabrese, F., Ratti, C.: Real time rome. Netw. Commun. Stud. 20(3-4), 247–258 (2006)
9. Beresford, A.R., Stajano, F.: Location privacy in pervasive computing. IEEE Pervasive Comput. 1, 46–55 (2003)
10. Konidala, D.M., Deng, R.H., Li, Y., Lau, H.C., Fienberg, S.E.: Anonymous authentication of visitors for mobile crowd sensing at amusement parks. In: Deng, R.H., Feng, T. (eds.) ISPEC 2013. LNCS, vol. 7863, pp. 174–188. Springer, Heidelberg (2013). doi: 10.1007/978-3-642-38033-4_13
11. Ganti, R.K., Fan, Y., Hui, L.: Mobile crowdsensing: current state and future challenges. IEEE Commun. Mag. 49(11), 32–39 (2011)
12. Lane, N., et al.: A survey of mobile phone sensing. IEEE Commun. Mag. 48(9), 140–150 (2010)
13. Bellavista, P., et al.: Scalable and cost-effective assignment of mobile crowdsensing tasks based on profiling trends and prediction: the ParticipAct living lab experience. Sensors 15(8), 18613–18640 (2015)
14. Namiot, D., Sneps-Sneppe, M.: On software standards for smart cities: API or DPI. In: ITU Kaleidoscope Academic Conference: Living in a converged world-Impossible without standards? Proceedings of the 2014. IEEE Press (2014)
15. Yue, K., et al.: Research of embedded database SQLite application in intelligent remote monitoring system. In: 2010 International Forum on Information Technology and Applications (IFITA), vol. 2. IEEE (2010)
16. Namiot, D., Sneps-Sneppe, M.: On open source mobile sensing. In: Balandin, S., Andreev, S., Koucheryavy, Y. (eds.) NEW2AN 2014. LNCS, vol. 8638, pp. 82–94. Springer, Cham (2014). doi:10.1007/978-3-319-10353-2_8
17. Funf. http://funf.org/. Accessed Jan 2017
18. Bansal, S.K.: Towards a semantic extract-transform-load (ETL) framework for big data integration. In: 2014 IEEE International Congress on Big Data. IEEE Press (2014)
19. Namiot, D.: On big data stream processing. Int. J. Open Inf. Technol. 3(8), 48–51 (2015)
20. Kroß, J., Brunnert, A., Prehofer, C., Runkler, T.A., Krcmar, H.: Stream processing on demand for lambda architectures. In: Beltrán, M., Knottenbelt, W., Bradley, J. (eds.) EPEW 2015. LNCS, vol. 9272, pp. 243–257. Springer, Cham (2015). doi:10.1007/978-3-319-23267-6_16
21. Lambda architecture. http://lambda-architecture.net/. Accessed Jan 2017
22. Questioning the Lambda Architecture. http://radar.oreilly.com/2014/07/questioning-the-lambda-architecture.htm. Accessed Jan 2017
23. Gál, Z., Hunor S., Béla G.: Information flow and complex event processing of the sensor network communication. In: 2015 Proceedings of the 6th IEEE International Conference on Cognitive Infocommunications (CogInfoCom). IEEE Press (2015)
24. Merging Batch and Stream Processing in a Post Lambda World. https://www.datanami.com/2016/06/01/merging-batch-streaming-post-lambda-world/. Accessed Jan 2017

25. Garg, N.: Apache Kafka. Packt Publishing Ltd, Birmingham (2013)
26. Maarala, A.I., et al.: Low latency analytics for streaming traffic data with Apache Spark. In: 2015 IEEE International Conference on Big Data (Big Data). IEEE Press (2015)
27. Kafka Streams. http://www.confluent.io/blog/introducing-kafka-streams-stream-processing-made-simple. Accessed Jan 2017
28. Apache Edgent. https://edgent.apache.org/. Accessed Jan 2017
29. Mobile-edge computing executing brief. https://portal.etsi.org/portals/0/tbpages/mec/docs/mec%20executive%20brief%20v1%2028-09-14.pdf. Accessed Jan 2017
30. Sanaei, Z., et al.: Heterogeneity in mobile cloud computing: taxonomy and open challenges. IEEE Commun. Surv. Tutorials **16**(1), 369–392 (2014)
31. Osseiran, Afif, et al.: Scenarios for 5G mobile and wireless communications: the vision of the METIS project. IEEE Commun. Mag. **52**(5), 26–35 (2014)
32. Namiot, D., Sneps-Sneppe, M.: On hyper-local web pages. In: Vishnevsky, V., Kozyrev, D. (eds.) DCCN 2015. CCIS, vol. 601, pp. 11–18. Springer, Cham (2016). doi: 10.1007/978-3-319-30843-2_2
33. Scalable Streaming of Video using Amazon Web Services. http://www.slideshare.net/AmazonWebServices/2013-1021scalablestreamingwebinar. Accessed Jan 2017
34. Selectel API (Russia). https://selectel.ru/services/cloud-storage/. Accessed Jan 2017
35. Apache CloudStack. https://cloudstack.apache.org/. Accessed Jan 2017
36. Nurmi, D., et al.: The eucalyptus open-source cloud-computing system. In: 2009 Proceedings of the 9th IEEE/ACM International Symposium on Cluster Computing and the Grid, CCGRID 2009. IEEE Press (2009)
37. OpenStack. https://www.openstack.org/. Accessed Jan 2017
38. Wen, X., et al.: Comparison of open-source cloud management platforms: OpenStack and OpenNebula. In: 2012 Proceedings of the 9th International Conference on Fuzzy Systems and Knowledge Discovery (FSKD). IEEE Press (2012)
39. IBM Cloud Video. https://www.ibm.com/cloud-computing/solutions/video/. Accessed Jan 2017
40. Smartvue. http://smartvue.com/cloud-services.html. Accessed Jan 2017
41. FI-WARE Cloud Hosting. https://forge.fiware.org/plugins/mediawiki/wiki/fiware/index.php/Cloud_Hosting_Architecture. Accessed Jan 2017
42. Kurento – the stream-oriented generic enabler. https://www.fiware.org/2014/07/04/kurento-the-stream-oriented-generic-enabler/. Accessed Jan 2017
43. Gheith, A., et al.: IBM bluemix mobile cloud services. IBM J. Res. Dev. **60**(2-3), 7:1 (2016)
44. Apache Usergrid. https://usergrid.apache.org/. Accessed Jan 2017
45. Rackspace Cloud Files. https://www.rackspace.com/cloud/files. Accessed Jan 2017
46. Namiot, D., Sneps-Sneppe, M.: On the domestic standards for Smart Cities. Int. J. Open Inf. Technol. **4**(7), 32–37 (2016)
47. Namiot, D., Sneps-Sneppe, M.: On physical web browser. In: Open Innovations Association and Seminar on Information Security and Protection of Information Technology (FRUCT-ISPIT), pp. 220–225. IEEE Press, New York (2016)

Research Infrastructures

The European Strategy in Research Infrastructures and Open Science Cloud

Konstantinos M. Giannoutakis[✉] and Dimitrios Tzovaras

Centre for Research and Technology Hellas, Information Technologies Institute,
57001 Thessaloniki, Greece
{kgiannou,Dimitrios.Tzovaras}@iti.gr

Abstract. The European Strategy Forum on Research Infrastructures (ESFRI) was established in 2002, with a mandate from the EU Council to support a coherent and strategy-led approach to policy-making on research infrastructures in Europe, and to facilitate multilateral initiatives leading to the better use and development of Research Infrastructures (RIs), at EU and international level. ESFRI has recently presented its updated 2016 Roadmap which demonstrates the dynamism of the European scientific community and the commitment of Member States to develop new research infrastructures at the European level.

Recently, the European Open Science Cloud (EOSC) initiated activities towards facilitating integration in the area of European e-Infrastructures and connected services between the member states, at the European level and internationally. It aims to accelerate and support the transition to an effective Open Science and Open Innovation in the Digital Single Market by enabling trusted access to services, systems and re-use of scientific data.

This work is focused on the identification of the new features and conclusions of the ESFRI Roadmap 2016 in terms of the methods and procedures that led to the call, the evaluation and selection of the new ESFRI Projects and the definition and assessment of the ESFRI Landmarks. An analysis of the impact of research infrastructures on structuring the European Research Area as well as the global research scene, and of the overall contribution to European competitiveness are also discussed. The EOSC challenges, purpose and initial recommendations for a preparatory phase that will lead to the establishment of the ambitious infrastructure for Open Science are also presented.

Keywords: Research infrastructures · European open science cloud · e-Infrastructures · e-Infrastructure reflection group

1 Introduction

ESFRI has recently fulfilled the commitment made by Member States and the European Commission in the Innovation Union flagship initiative and has implemented 60% of ESFRI projects by the end of 2015. The new ESFRI 2016 Roadmap demonstrates the dynamism of the European scientific community and the commitment of Member States to develop new research infrastructures at the European level. The networks of research infrastructures across Europe strengthen its human capital base by providing world-class

L. Kalinichenko et al. (Eds.): DAMDID/RCDL 2016, CCIS 706, pp. 207–221, 2017.
DOI: 10.1007/978-3-319-57135-5_15

training for a new generation of researchers and engineers and promoting interdisciplinary collaboration.

The objective of Horizon 2020 is to ensure the implementation and operation of the ESFRI and other world class research infrastructures, including the development of regional partner facilities; integration of and access to national research infrastructures; and the development, deployment and operation of e-infrastructures. The major changes in Horizon 2020 with regard to the previous framework programme (FP7) include the emphasis on innovation and the development of human resources, addressing industry as an e-infrastructure supplier and user, more resolve towards service orientation and service integration, and more emphasis on data infrastructure development. In Horizon 2020 the e-infrastructure activities are part of the European Research Infrastructures, including e-infrastructures programme. The indicative budget for e-infrastructures from 2014 to 2020 is 890 million euros.

This paper starts with an overview of the European strategy for RI, with a special emphasis in e-Infrastructures, as defined in the ESFRI Strategy report published in March 2016 and will then show that implementations of some of the facilities stipulated by the Strategy have already started under the H2020 support. Focus is given on the identification of the new features and conclusions of the ESFRI Roadmap 2016 in terms of the methods and procedures that led to the call, the evaluation and selection of the new ESFRI Projects and the definition and assessment of the ESFRI Landmarks. An analysis of the impact of research infrastructures on structuring the European Research Area as well as the global research scene, and of the overall contribution to European competitiveness will also follow. A Landscape Analysis will also be presented that provides the current context, in each domain, of the operational national and international research infrastructures open to European scientists and technology developers through peer-review of competitive science proposals. The e-infrastructures landscape, transversal to all domains, will be also elaborated as approached by the e-Infrastructure Reflection Group (e-IRG).

The work also focuses on recent initiatives and activities supporting the e-infrastructure activities in Horizon 2020 in order to achieve by 2020 a single and open European space for on-line research where researchers will enjoy leading-edge, ubiquitous and reliable services for networking and computing, and seamless and open access to e-Science environments and global data resources. These initiatives are:

- The **European Open Science Cloud** initiative activities, towards facilitating integration in the area of European e-Infrastructures and connected services between the member states, at the European level, and internationally.
- Activities of the **e-Infrastructure Reflection Group (e-IRG),** focusing on presenting the e-IRG Roadmap 2016, published in December 2016, which gives guidance and recommendations for policy and technical discussions on the main European Open Science Cloud topics.

Ultimately, the guidelines for providing access to RIs are given, and the impact of RIs and the EOSC to European Union and internationally are discussed. It should be noted that this paper extends the work of [9] by introducing the European Open Science Cloud first report, which outlines initial recommendations for the realization of the Open

Science Cloud in Europe, and the updated e-IRG Roadmap 2016 that proposes the recommendations for proper integration of European e-Infrastructures and connected services between Member States.

2 European Strategy for Research Infrastructures

Since 2006, ESFRI focuses on the identification of RIs in Europe across all scientific areas and periodically updates its roadmaps for providing a strategic vision for ensuring the access of RIs to researchers. Generally, the main objectives of ESFRI, as described in [6], are:

- to support a coherent and strategy-led approach to policy making on research infrastructures in Europe
- to facilitate multilateral initiatives leading to a better use and development of research infrastructures acting as an incubator for pan-European and global research infrastructures
- to establish a European Roadmap for research infrastructures (new and major upgrades, pan-European interest) for the coming 10–20 years, stimulate the implementation of these facilities, and update the Roadmap as the need arises
- to ensure the follow-up of implementation of already ongoing ESFRI projects after a comprehensive assessment, as well as the prioritization of the infrastructure projects listed in the ESFRI Roadmap.

With the last roadmap, published in 2016, ESFRI has widened its horizon and scope by adopting a more focused, strategic approach and identifying a limited number of RIs with high added value for European research. The target is to fund a limited number of mature projects that will enhance European research and innovation competitiveness. Thus, as stated in [6], ESFRI added as an important eligibility condition that a proposal required a funding commitment from the submitting Member State or Associated Country along with a political commitment from at least two others. This requirement strengthened the transparency of the submission process and forced a dialogue and crosscheck between the research communities and the concerned governments from the very beginning of the RI project, [6]. This is also ensured by performing a landscape analysis of RIs for identifying their strengths, potential and weaknesses in all fields of research.

2.1 Landscape Analysis and ESFRI Landmarks

The purpose of landscape analysis is to identify the operating open access RIs from national, regional and international infrastructures, as well as groups that provide integrated solutions with open access to the state-of-the-art resources. The impact of the landscape analysis by the ESFRI infrastructures is emphasized by the list of ESFRI Landmarks, that are implemented ESFRI projects (or started implementation under the roadmap) with great success on providing scientific services and competitiveness of the European Research Area.

The key elements of the new ESFRI process, as stated in [6], are:

- definition of clear rules, communication and explanation of the procedure at the start;
- delineation of a window of opportunity: new projects will remain on ESFRI Roadmap for a maximum of ten years;
- evaluation of scientific relevance and project maturity in parallel but separately;
- engagement of international experts and peer reviewers in the evaluation process;
- adoption of a lifecycle approach to the analysis of infrastructures, with *Projects and Landmarks* clearly identified and indication of *emerging opportunities;*
- assessment of the implementation of the inherited projects from Roadmap 2008 and 2010 for monitoring their progress and identifying areas where support is needed;
- recognition and analysis of the overall "Landscape" of the European research infrastructure system and of the complementarity of projects;
- identification of the role of the successful ESFRI infrastructures and definition of the "Landmark list";
- monitoring of Projects and periodic review of Landmarks, and update of the Roadmap.

Emphasis on new ESFRI projects is always given on excellence, impact, sustainability and continuous report.

2.2   The ESFRI Roadmap 2016 Projects and Landmarks

The new ESFRI roadmap contains 21 ESFRI projects, 9 from the 2008 roadmap, 6 from the 2010 roadmap and five new project (plus one reoriented project). The evaluation process for selecting the new projects was (a) the Strategy Working Groups with respect to their scientific excellence, pan-European relevance and socio-economic impact and (b) their degree of maturity as benchmarked by the ESFRI Implementation Group.

Regarding the ESFRI Landmarks, 29 are listed containing already implemented projects and two new projects that were evaluated and are under construction. The list of the ESFRI projects and the ESFRI Landmarks are presented in [6], and are categorized in the following application categories: Energy (4 projects), Environment (5 projects), Health & Food (8 projects), Physical Sciences & Engineering (3 projects), and Social & Cultural Innovation (1 project).

2.3   Big Data in Research Infrastructures

Data occurring from RIs by experimentation, measurements, observations, data analysis, modelling and simulations are usually large or complex that traditional data processing applications are insufficient to process. Moreover, the analysis of such data is crucial to scientific research and usually require high performance computing or cloud computing capabilities in order to process them.

For this reason, the Research Data Alliance (RDA) addresses a global data policy with direct connections with e-IRG, with specific actions on standardization strategies, like PanData for analytic facilities, [8]. Usually, such data occur from physics experiments, astronomical research, biomedical interests or large scale simulations of complex systems. The distributed nature of RIs in the domains listed before, enhances the need for effective data access and analysis. Thus, the ESFRI is expected to play a significant role in this general development, by formulating a new generation of big data practitioners and big data engineers.

2.4 ESFRI Evaluation Process

The evaluation process of new projects from ESFRI adopted a transparent approach together with national research authorities and research communities of the new roadmap process. This process is illustrated in Fig. 1, and consists of five distinct steps, as presented in [6]:

- *Review of projects from previous ESFRI roadmaps*. The RIs that were implemented and produced high quality services to the community led to the identification of the ESFRI Landmarks. The lifecycle analysis performed for the identification of scientific key services is now adopted and key issues can now be resolved for the remaining projects.
- *Submission and eligibility of proposals*. Members or Associated countries can submit proposals to the ESFRI Roadmap 2016. New ESFRI Projects have to be competitive and mature, while the ability to be implemented within ten years is crucial. They need to demonstrate government level financial commitment of the proponent Member State or Associated Country plus at least two additional political commitments. This increases the likelihood of success for the projects and enables a more robust and reliable selection process by ESFRI.
- *Evaluation of proposals*. Eligible proposals were assessed through two parallel and independent evaluation processes. The Strategy Working Groups (SWGs) evaluated the scientific case, i.e. scientific merit, relevance and impact, European added value, socio-economic benefit and the needs of interfacing or integrating external e-infrastructure. The SWG identified assigns a minimum of three independent international peer-reviewers who contribute their evaluation on the science aspects of the project. In parallel, the Implementation Group (IG) assessed the maturity, i.e. stakeholder commitment, user strategy and access policy, preparatory work, planning, governance and management, human resources policy, finances, feasibility and risks. The IG similarly assigns international expert evaluators to assess the relevant "maturity" aspects of each project. Based on their own analysis and on the reports from the referees, the SWG and IG identified critical questions and issues to be addressed by each of the eligible proposals. The SWG and IG subsequently reached their conclusions with a joint recommendation per project, and an overall harmonization to align the results from the different areas and formulate a ranking of the projects and recommendations.
- *Decision making*. The executive board proposes a final recommendation on the list of projects and Landmarks that should be included for a final decision.

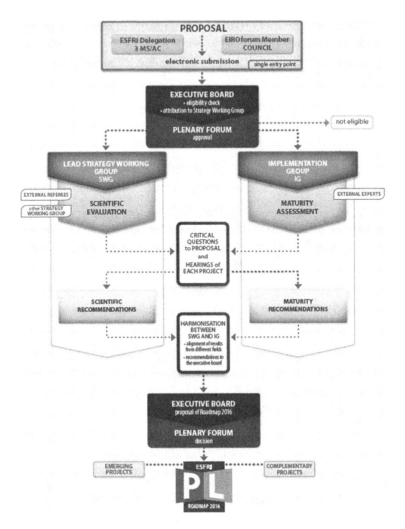

Fig. 1. The ESFRI evaluation process of new projects, [6].

3   e-Infrastructure Activities in Horizon 2020

During the last decades, scientific applications require challenging demands of interoperable data, computational power and collaborations between different scientific fields and researchers. The developed large scale information systems called e-infrastructures and are supported by many initiatives in Europe and worldwide.

The European Commission (EC) is interested on investing through its Framework Programmes in e-Infrastructures since they are considered as key enablers of the European Research Area (ERA). The collaboration among scientific communities of researchers that work together on complex multi-disciplinary problems whose solutions are highly beneficial for the society and the progress at large are then of high importance, [1].

The corresponding H2020 call for e-infrastructures (H2020-EINFRA-2016–2017) focuses on the open research data, data and computing intensive science, research and education networking, high performance computing and big data innovation, [4]. The following principles are key elements for the project implementations:

- *Service orientation*: The funded projects will contribute to the formulation of a high quality catalogue of services describing the services that they will provide during the lifetime of the projects. A service oriented European e-infrastructure landscape will be supported by adopting new knowledge and innovative ICT solutions by global and multidisciplinary research.
- *Maximizing and assessing the impact of the e*-infrastructures: Projects should have a clear plan for active participation in international fora and research groups/initiatives to promote data and computing infrastructure interoperability.
- *Co-design*: A balanced set of partners with complementary competences and roles should form the consortium of projects that will be funded by this call.
- *Open Research Data*: In order to make research data discoverable, accessible, intelligible, useable and interoperable, this call focuses on open data taking into account the above criteria.
- *H2020 as a catalyst of the European plan for growth and jobs*: This call will promote the use of other funding sources as instrument to support initiatives of European interest to foster growth and jobs.

The call has two themes: *"Integration and consolidation of e-infrastructure platforms supporting European policies and research and education communities"* and *"Support to the next implementation phase of Pan-European High Performance Computing infrastructure and services (PRACE)"* with several topics. More details and the specific topics can be found in [3].

Recent initiatives and activities for supporting the e-infrastructure activities in Horizon 2020 have been formulated. In order to achieve by 2020 a single and open European space for on-line research where researchers will enjoy leading-edge, ubiquitous and reliable services for networking and computing, and seamless and open access to e-Science environments and global data resources, several research groups and cooperation have been instantiated. The distribution of Research Infrastructures in Europe is illustrated in Fig. 2.

4   European Open Science Cloud

Recently, the Commission High Level Expert Group on the European Open Science Cloud (HLEG EOSC) has drafted a first report for the realization of the European Open Science Cloud. With this initiative, the European Commission with the EOSC aims to introduce, accelerate and support the Open Science and Open Innovation in the Digital Single Market, thus enabling re-use of geographically distributed scientific data.

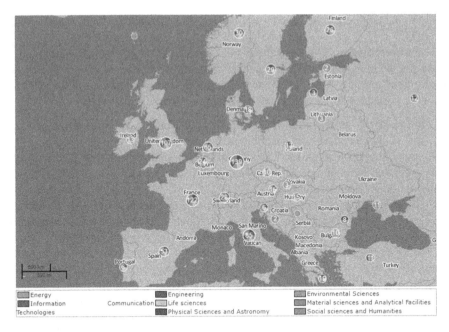

Fig. 2. Research Infrastructures funded by the European Commission (http://ec.europa.eu/research/infrastructures/index_en.cfm?pg=mapri).

The main purpose of EOSC is the re-use of research data and accompanying tools, the access support to the provided services and systems without any social and geographical borders. The professional data management and long term data maintenance is of main concern and is expected to enable the re-use and data driven knowledge discovery and innovation.

It has been reported that approximately 50% of the produced research data are not reproducible, while more than 80% of data are not stored into a secure and sustainable repository, [5]. Since many scientific areas produce huge amount of data especially during the last years, [5], the need for data stewardship and a global data and services framework is of great importance. For example, DNA sequence data is doubling every 6–8 months over the last 3 years and this rate looks to be steady for this decade.

The main purpose of EOSC is to remove all the technical, governmental and human barriers in order to accelerate the re-use of research data and to support the access to services without any social or geographical borders. The main elements for success for the EOSC have been outlined in [2], and are:

- *Open*: EOSC should be open in design, participation and in use. Its open access will promote the development and adoption of open standards, thus enabling collaborative environments with no barriers to participation or resource-sharing. It will enable accessibility, transparency and reproducibility in all stages of the research life cycle. EOSC will foster public-private partnerships with target to turn all investments into economic growth.

- *Publicly funded and governed*: A publicly funded and governed Open Science Cloud will guarantee persistence and sustainability, while ensuring that outcomes are driven by scientific excellence and societal needs. The partnership with private sector actors will encourage the development of innovative services that are conducive to the future Open Science, while guaranteeing the long term and persistent care of resources.
- *Research-centric*: Research community will be engaged in the design of the Open Science Cloud to ensure the development of services according to their needs.
- *Comprehensive*: EOSC will not target a specific scientific discipline or research field. It will promote inter and multi-disciplinary science and encourage innovation and integrated knowledge creation and sharing among all research communities.
- *Diverse and distributed*: EOSC will utilize the Europe's distributed e-Infrastructures for forming a network of actors, resources and services organized nationally and at European level. It will drive a more efficient use of ICT investments across infrastructures and communities, lowering the barriers to adoption for institutions and researchers.
- *Interoperable*: With the adoption of common open standards and protocols for all resources and digital services, the Open Science Cloud will connect the corresponding networks, data, computing systems, tools and services for research with an interoperable way.
- *Service oriented*: It will provide services that address the full research lifecycle, including data gathering, management, analysis, sharing and discovery. The EOSC will be the framework for new innovative methodologies and services that further advance research in the Open Science context.
- *Social*: EOSC will be a socio-technical endeavor that connects diverse communities and promotes the development of human networks. It will enable the sharing of knowledge and facilitate the embedding of Open Science practices into researcher's workflows.

During the preparatory phase, the High Level Expert Group drafted the following recommendations. More details can be found in [5].

- *Policy recommendations*
  - *Take immediate, affirmative action on the EOSC in close concert with Member States*. Member states should take immediate actions to realize the first phase of a federated, globally accessible environment, where researchers can publish and re-use data and tools for research, innovation and educational purposes.
  - *Close discussions about the 'perceived need'*. The preparatory phase should not be long, since there has been a long consultation phase while a variety of services and data are already available.
  - *Build on existing capacity and expertise where possible*. The majority of the services and data already exist as parts of e-Infrastructures and are of high quality. Therefore, it is believed that this preparatory phase will be significantly progressed or completed by the end of 2017.
  - *Frame the EOSC as the EU contribution to an Internet of FAIR Data and Services underpinned with open protocols*. This will allow open and common

implementation together with participation from all stakeholders, including research infrastructure providers, Member States, research institutes and businesses. All providers can start implementing prototype applications for the Internet of FAIR data and Services on the day minimal standards and the minimal rules of engagement are released.

- *Governance recommendations*
  - *Aim at the lightest possible, internationally effective governance.* An inclusive, flexible, transparent and decentralized approach is required in order to establish a lightweight, sustainable and collaborative governance model for the EOSC and its stakeholders.
  - *Guidance only where guidance is.* Due to the large number of expected stakeholders, guidance will be offered for harmonization and sustainability purposes.
  - *Define Rules of Engagement for service provision in the EOSC.* The development of the infrastructure will be guided and governed by a minimal set of protocols and rules of engagement that specify the conditions under which stakeholders can participate.
  - *Federate the gems and amplify good practice.* Based on the consensus that most foundational building blocks of the Internet of FAIR data and Services are operational somewhere, it is recommended that early action is taken to federate these gems. Optimal engagement is required of the e-infrastructure communities, the ESFRI communities and other disciplinary groups and institutes.
- *Implementation recommendations*
  - *Turn the HLEG report into a high-level guide to scope and guide the EOSC initiative.* The first HLEG report should serve as a high level guiding document for the actual development and implementation processes in the Member States and in Horizon 2020 work programme.
  - *Develop, endorse and implement the Rules of Engagement for the EOSC.* The Commission should develop the Rules of Engagement (RoE) for stakeholders that want to provide a component of the EOSC. RoE should be based on the assumption that all data in EOSC are FAIR.
  - *Set initial guiding principles to kick-start the initiative as quickly as possible.* Specific principles have been proposed in order to guide the preparatory phase.
  - *Fund a concerted effort to develop core data expertise in Europe.* A substantial training initiative in Europe is proposed, to locate, create, maintain and sustain the required data expertise. The aim of this training is to train certified core data experts, consolidate and develop assisting material and tools for the construction of data management plans and data stewardship plans, and by 2020 to have in each Member State at least one certified institute to support implementation of Data Stewardship per discipline.
  - *Develop a concrete plan for the architecture of data interoperability of the EOSC.* Concrete actions for Member States are proposed that will ensure data interoperability.
  - *Install an innovative guided funding scheme for the preparatory phase.* For the preparatory phase of EOSC Horizon 2020 funds will be used, but a new funding

mechanism will be established, specifically designed to rapidly prototype, test and reach all goals.

- *Make adequate data stewardship mandatory for all research proposals.* Horizon 2020 projects with achievable requirements, that are data multidisciplinary, that properly address post-project sustainability or otherwise advance the common aims of the EOSC should be streamlined for funding in the EOSC game-changer scheme.
- *Provide a clear operational timeline to deal with the early preparatory phase of the EOSC.* Concrete tasks have been identified to kick start the EOSC: (1) Define the RoE, (2) Create and foster cross-domain collaboration including ESFRIs and e-Infrastructures, (3) Define training needs and models for the training infrastructure, (4) Develop a governance plan for EOSC, (5) Establish minimal technical standards and plan for their long-term maintenance and compliance, (6) Establish guidance and oversight mechanisms for the EOSC fame changer scheme, and (7) Establish plans for certified institutes for data expertise and stewardship.

The EOSC reflects the change in the way research is conducted but funding and the long-term perspective is missing. The preparatory phase needs to look into governance, additional funding and sustainability. Incentives are needed for researchers to share and re-use their data and this is closely tied to the quality of services which are key for the uptake and success of the platform. The EOSC must be built upon existing e- and research infrastructures, and it should be a mixed bag of infrastructures, tools and services presented as interoperable virtual environments, "a cloud of services", for all European researchers to store, manage, process, analyze and re-use research data across geographical, social and technical borders.

4.1   e-Infrastructure Reflection Group (E-IRG)

The e-Infrastructure Reflection Group (e-IRG), founded in 2003, is a European independent body with mission to form a general purpose European e-Infrastructure by providing guidance on the development of a European e-Infrastructure for science and research. Some examples of already established e-Infrastructures across Europe are GEANT (networking - http://www.geant.org/), EGI (grid and cloud computing - https://www.egi.eu/), PRACE (supercomputing - http://www.prace-ri.eu/), EUDAT (European Data Infrastructure - https://www.eudat.eu/), IDGF (crowd computing - http://idgf-sp.eu/), OpenAIRE (repository for scientific articles - https://www.openaire.eu/) and LIBER (Association of European Research Libraries - http://libereurope.eu/).

After the publication of the ESFRI Roadmap 2016, the e-IRG group released the e-IRG Roadmap 2016 during the December of 2016, which define a clear route on how to evolve the European e-Infrastructure system further, [7]. The e-IRG group uses the term e-Infrastructure Commons for the e-Infrastructure resources and related services (originally this term was presented in the e-IRG Roadmap of 2012), to target its realization by 2020. e-IRG states that the implementation of the e-Infrastructure Commons will be a large step towards the European leadership in research infrastructures including e-Infrastructures, the realization of the European Open Science Cloud and the European Data Infrastructure, [7].

The Roadmap proposes some key recommendations, [7], which are briefly:

- RIs and research communities should reinforce their efforts to: elaborate on and drive the e-Infrastructure needs; participate in the innovation of e-Infrastructure services; contribute to standards and take care of their data.
- e-Infrastructure providers should further increase efforts to work closely together to fulfil the complex user needs in a seamless way.
- National governments and funding agencies should reinforce their efforts to: embrace e-Infrastructure coordination at the national level and build strong national e-Infrastructure building blocks, enabling coherent and efficient participation in European efforts; Analyze and evaluate their national e-Infrastructure funding and governance mechanisms, identify best practices and provide input to the development of the European e-Infrastructure landscape.
- European Commission should provide strong incentives for cross-platform innovations and further support the coordination and consolidation of e-Infrastructure service development and provisioning on the national and European level.

Generally, e-Infrastructures can be considered as an essential building block of the European Research Area, while the e-Infrastructure Commons aims to be an essential building block for the EOSC. Representatives from the e-Infrastructures EUDAT, LIBER, OpenAIRE, EGI and GEANT proposed their view of the realization of an e-Infrastructure Commons in [2], where they state that the majority of the needed EOSC services already exists and only technical and policy barriers remain.

5   Access to Research Infrastructures

European Commission, in close cooperation with the ESFRI, the e-IRG group and other EU organizations developed the Charter for Access to Research Infrastructures, for promoting the harmonisation of access procedures and enhanced transparency of access policies in order to enable the remote access of users to the RIs. This document was published in March 2016, [3], and it is accompanied with reference documentation containing complementary material regarding the definition of an access policy for any RI.

The charter promotes access to RIs and interaction with a wide range of social and economic activities, including business, industry and public services, in order to maximise the return on investment in RIs and to drive innovation, competitiveness and efficiency, [3]. The Charter for Access to RIs proposes the following guidelines that each RI have to specify, [3]:

- *Access policy*: The Access Policy should define the access in terms of Access Units, the state of the specific Access mode, clarify the conditions for Access, describe the processes and interactions involved in the Access and elaborate on the support measures facilitating the Access.

- *Access modes*: Three different Access modes have been defined, i.e. excellence-driven, market-driven and wide. Thus, each access to a RI may be regulated according to one Access mode or any combination of them.
- *Access restrictions*: Definition of possible restrictions by means of quota or pre-defined user groups.
- *Access processes and interactions*: The following processes and interactions are defined in the Access to RIs: application, negotiation, evaluation, feedback, selection, admission, approval, feasibility check, setting-up, use, monitoring and dismantling.
- *Support measures facilitating Access*: RIs are encouraged to offer support measures such as guidance through user manuals, provision of user support, provision of accommodation and guidance with immigration procedures.
- *Education and training*: RIs are encouraged to offer education and training, as well as to collaborate with other institutions and organizations that benefit from using RIs for their education and training purposes.
- *Regulatory framework*: A regulatory framework should be defined when providing access to a RI, that should cover Access, intellectual property rights, data protection, confidentiality, liability and possible fees.
- *Transparency*: Each RI should provide transparent information on the RI itself, including its services, access policy, data management policy and the terms and conditions.
- *Research data management plan*: RIs and users should have an agreement on a data management plan outlining how the research data will be handled.
- *Health, safety, security and environment*: RIs should take the necessary actions to ensure the health, security and safety of any user accessing the RI itself, as well as to minimize the impact on the environment.
- *Quality assurance*: RIs are encouraged to set up mechanisms in order to evaluate the quality of the provided access to users.
- *Limitations*: Access to RIs may by limited by the following: national security and defence; privacy and confidentiality; commercial sensitivity and intellectual property rights; ethical considerations in accordance with applicable laws and regulations.

Therefore, all the established RIs have published[1] in accordance with the charter, their individual access policies to foster collaboration among researchers across Europe. International cooperation and access to RIs is strongly encouraged from the European Commission, thus RIs should also define how non-EU members can grand access to their infrastructures.

6   Impact of Research Infrastructures and EOSC

The relationship between academia and RIs contributes to an effective European educational and scientific ecosystem that attracts and supports industry. The optimal distribution of the RIs across Europe (Fig. 2), is of great importance since it contributes in promoting European cohesion, [6]. The knowledge-based economy is effectively

---

[1] https://ec.europa.eu/research/infrastructures/index_en.cfm?pg=charter_access_ri.

stimulated by strengthening the links between RIs, higher education and research institutions with economic players like industry, services and utilities, [6].

The ESFRI Working Group on Innovation (WG INNO) promoted the industrial capabilities of the RIs on the ESFRI Roadmap in order to push the cooperation of pan-European RIs with industry. This implies promoting partnerships on R&D projects for realizing ecosystems of integrated competences, services and technologies facilitating industrial innovation. Generally, linking the RIs among themselves and allowing the broader social, technological and economical players to acquire information from multiple resources in an effective, efficient and sustainable way, [6].

The success of many RIs leads the G8+5 Group of Senior Officials (GSO) to define a strategy for Global Research Infrastructures (GRI). A well-known example of a GRI is CERN for high-energy physics that is now considered as a globalized infrastructure. The Global Science Forum is responsible for identifying GRI needs and opportunities, including the organizational and long term sustainability aspects. ESFRI acts as a reliable partner at the global level in the practical development of scientific and political initiatives aimed at internationalization of new or existing infrastructures that appear ready to move to a global operation involving access, data policy, and lifecycle management and to consider international governance, [6].

The Commission High Level Expert Group on the European Open Science Cloud expects that when the European Open Science Cloud and the European Data Infrastructure (EUDAT) will be accessible, it will bring benefits for:

- Businesses that will have cost-effective and easy access to top level data and computing infrastructure, as well as a wealth of scientific data enabling data-driven innovation. This will particularly benefit SMEs, which typically lack access to such resources.
- Industry will benefit from the creation of a large-scale cloud eco-system, support-ing the development of new European technologies such as low-power chips for high performance computing.
- Public services will benefit from reliable access to powerful computing resources and the creation of a platform to open their data and services, which can lead to cheaper, better and faster interconnected public services. Researchers will also benefit from online access to the wealth of data created by public services.

Making research data openly available can help boost Europe's competitiveness by benefitting start-ups, SMEs and data-driven innovation, including in the fields of medicine and public health.

7   Conclusions

European Commission and the ESFRI group have provided the resources and competence for supporting a strong research and innovation profile for their members, by sharing expensive scientific equipment and e-infrastructures, capitalising on cross-border collaboration and human potential across Europe. The main principles for the European strategy for Research Infrastructures rely on the adoption of a coherent participation model in European and global RI initiatives, utilization of a continuous collaboration among academia and industry, and on the harmonization of investment in e-infrastructures, as key enablers of a knowledge intensive economy, for and of eScience. A coordinated policy framework of e-Infrastructures, including fast networking, storage, high performance computing, data access and management structures and services has been progressing during the last years.

ESFRI, over the past decade, has improved the efficiency and impact of the European RIs, that are moving towards a sustainable investment for overall competitiveness. With the new published ESFRI Roadmap 2016, new methods and procedures have been adopted, as well as new procedures for the evaluation and selection of new projects and definition and assessment of the ESFRI Landmarks. Enhancing and optimising RIs and their access by scientists and researchers is a key element of competitiveness as well as a necessary basis for dealing with societal challenges. On the other hand, the EOSC initiative puts effort in order to start up a trusted open environment for storing, sharing and re-using scientific data and results supporting Open Science practices. This "cloud for scientists" will set up the standards for the management, interoperability and quality of scientific data that will promote public-private innovation to satisfy the needs of the research communities and increase the global competitiveness of European ICT providers.

References

1. Andronico, G., Ardizzone, V., Barbera, R., Becker, B., Bruno, R., Calanducci, A., Carvalho, D., Ciuffo, L., Fargetta, M., Giorgio, E., Rocca, G.L., Masoni, A., Paganoni, M., Ruggieri, F., Scardaci, D.: e-Infrastructures for e-Science: a global view. J. Grid Comput. 9(2), 155–184 (2011)
2. EUDAT, LIBER, OpenAIRE, EGI, GEANT. The European Open Science Cloud for Research, 30 October 2015. https://www.eudat.eu/sites/default/files/EOSC%20Paper%20Press%20Release-30%2010%202015.pdf
3. European Commission: European Charter for Access to Research Infrastructures. Principles and Guidelines for Access and Related Services. https://ec.europa.eu/research/infrastructures/pdf/2016_charterforaccessto-ris.pdf
4. European Commission: HORIZON 2020-Work Programme 2016–2017. European Research Infrastructures (including e-Infrastructures). http://ec.europa.eu/research/participants/data/ref/h2020/wp/2016_2017/main/h2020-wp1617-infrastructures_en.pdf
5. European Commission: Realising the European Open Science Cloud, First report and recommendations of the Commission High Level Expert Group on the European Open Science Cloud, Brussels (2016). http://ec.europa.eu/research/openscience/pdf/realising_the_european_open_science_cloud_2016.pdf
6. European Strategy Forum on Research Infrastructures: Strategy Report on Research Infrastructures, Roadmap (2016). http://www.esfri.eu/roadmap-2016
7. e-Infrastructure Reflection Group (e-IRG). E-IRG Roadmap (2016). http://e-irg.eu/documents/10920/12353/Roadmap+2016.pdf
8. Photon and Neutron data infrastructure initiative (PaNdata). http://pan-data.eu
9. Tzovaras, D.: Overview of the European strategy in research infrastructures. In: XVIII International Conference Data Analytics and Management in Data Intensive Domains (DAMDID/RCDL'2016), pp. 187–194 (2016). http://ceur-ws.org/Vol-1752/paper31.pdf

Creating Inorganic Chemistry Data Infrastructure for Materials Science Specialists

Nadezhda N. Kiselyova[1(✉)] and Victor A. Dudarev[1,2]

[1] A. A. Baikov Institute of Metallurgy and Materials Science of Russian Academy of Sciences (IMET RAS), Moscow, Russia
{kis,vic}@imet.ac.ru
[2] National Research University Higher School of Economics (NRU HSE), Moscow, Russia

Abstract. The analysis of the large infrastructure projects of information support of specialists realized in the world in the field of materials science is carried out (MGI, MDF, NoMaD, etc.). The brief summary of the Russian information resources in the field of inorganic chemistry and materials science is given. The project of infrastructure for providing the Russian specialists with data in this area is proposed.

Keywords: Information support in materials science · Materials science database integration · Inorganic substances and materials

1 Introduction

Global market competitive requirements demand permanent improvement and enhancing of consumer products properties. Products quality and novelty are substantially defined by the materials used in its production. As a result, acceleration of search, research and production of new materials with required functional properties is a crucial problem of industry and economic development for all countries in general. At present, according to the American specialists [1], the time frame for incorporating new classes of materials into applications is about 20 years from initial research to first use. It is connected with the fact that very often consumers have no sufficient information even about very promising materials, investigations in a research and technology development for materials preparation and processing are unreasonably duplicated, therefore substances and materials with not the best consumer and other parameters are used that leads to loss of products quality, production cost escalation and, eventually, to loss of the released product market attractiveness.

One of ways to accelerate new materials search, development and deployment is the mature infrastructure for specialists' information support creation. First of all, it is a creation of distributed virtually integrated network of the databases and knowledge bases containing information on properties of substances and materials and technologies of their production and processing, and also, it's development of systems for computer-aided design and modeling of the materials available from the Internet to wide range of

L. Kalinichenko et al. (Eds.): DAMDID/RCDL 2016, CCIS 706, pp. 222–236, 2017.
DOI: 10.1007/978-3-319-57135-5_16

specialists: to scientists, engineers, technologists, businessmen, government employees, students, etc.

In recent years in the developed countries the initiatives aimed to the infrastructure organization for access to experimental and calculated data about materials were announced and supported by the governments. The brief summary of some initiatives was given in some publication earlier [2, 3]. Current review considers set-theory approach to inorganic substances and materials database integration and the project of infrastructure for providing Russian specialists with data in the field of inorganic chemistry and materials science in more details comparing with [2].

2   Materials Genome Initiative (MGI)

In 2011 the USA started a project, called Materials Genome Initiative (MGI) [4]. The MGI aims are to provide accelerated creation of the new materials with a set of predefined properties which is critical for achievement of the high level of competitiveness of the industry in the USA and will promote support to their leading role in many modern materials science and industry areas: from power engineering to electronics, from national defense to health care. In MGI special attention is paid to breakthrough researches support in the theory, materials properties modeling and data mining as means of significant progress achievement in materials science that will lead to cost reduction during new materials research, development, and production. The MGI tasks are in ensuring new materials development and deployment, including research activities coordination and providing access to computable models and tools for materials properties and behavior assessment, and also breakthrough methods of modeling and data analysis usage. The MGI project implementation will allow creation of the mechanisms promoting materials data and knowledge exchange not only between researchers but also between the academic science and the industry. A basis of the MGI is Materials Innovation Infrastructure (MII) which provides integration of modern modeling methods and experimental research. The infrastructure includes interconnected service structures and objects complex (including the objects of megascience) making and/or providing a basis for materials science functioning as a science and in the applied area. Subcommittee on the MGI The National Science and Technology Council (NSTC) includes representatives of the United States Department of Defense, Department of Energy, National Institute of Standards and Technology (NIST), National Science Foundation (NSF), National Aeronautics and Space Administration (NASA), National Institutes of Health (NIH), United States Geological Survey (USGS), Defense Advanced Research Projects Agency (DARPA), etc. [5]. The successful AFLOW system [6, 7] should be highlighted among MGI-supported projects. It contains a database with substances quantum mechanical calculations results and equipped with the computer software package for carrying out such calculations. The new type of the ultra-stable and wearproof glass discovery was also performed by wide use of theoretical calculations [8] within MGI. Another example is the virtual high-throughput experimentation facility with the goal of accelerating the generation of huge data volumes, needed to validate existing materials models for new substances with predefined properties prediction (NIST and the National Renewable Energy Laboratory (NREL)), etc. [9].

3   The Materials Data Facility (MDF)

Considering materials importance for the US industry high level competitiveness achievement, in June, 2014 the National Data Service (NDS) Consortium announced a pilot project for materials science data facilities development: The Materials Data Facility (MDF) [10] supported by NIST. This project is the answer to the MGI of the White House aimed to accelerate modern materials development. The MDF will provide materials scientists with a scalable repository for experimental and calculated data (including prior to their publication) storage, supplied with references to the appropriate bibliographic sources. The MDF will be an instrument of national infrastructure creation for collective information use, including DBs on materials properties developed in the world and information systems for calculation and modeling, and also will promote materials data exchange, including not published data. Materials data and calculation facilities availability is provided by means modern information and telecommunication infrastructure which allows to provide data to materials researchers for multi-purpose use, additional analysis and verification. In addition to NIST, it is necessary to note several main contributors to MDF: The University of Chicago, Argonne National Laboratory, The University of Illinois, Northwestern University, Center for Hierarchical Materials Design, etc. The MDF repository currently includes [10], in addition to a numerous NIST DBs [11], information systems with quantum mechanical calculations results: AFLOW [6, 7], The Open Quantum Materials Database (OQMD) [12], etc.

4   Novel Materials Discovery Laboratory (NoMaD)

The NoMaD project was the European Union answer to the US MGI. The NoMaD project [13, 14] is directed to the European Centers of Excellence creation and contemplates development of DB network (Materials Encyclopedia) on substances and materials properties (mainly on the calculations results), and also a number of facilities for data analysis and substances calculation. The purpose is materials with predefined functional properties development and use acceleration. The program started in November, 2015 within the EU's HORIZON2020 project [14]. Essential disadvantage of the NoMaD is an orientation to the US information resources (mainly, NIST DBs on substances and materials properties) and information systems with calculated data. Nowadays the NoMaD repository [15] contains already synthesized substances quantum mechanical calculations results only. In many respects the NoMaD program correlates with the EU Materials design at the eXascale (MaX) [16] project including infrastructure creation for carrying out quantum mechanical calculations by means of high-performance computer systems. Among the NoMaD partners there are a number of Europe's leading organizations, such as Humboldt University, Fritz-Haber-Institute of the Max Planck Society, King's College London, University of Barcelona, Aaalto University, Max Planck Institute for the Structure of Dynamics of Matter, Technical University of

Denmark, Max Planck Computing and Data Facility, Barcelona Supercomputing Centre, etc.

5 Materials Research by Information Integration Initiative (MI²I)

The MI²I was offered in 2015 by the Japanese government, which created the Center for Materials Research by Information Integration based on the National Institute for Materials Science (NIMS) [17]. Unlike the European programs the created center is aimed to wide use of not only quantum mechanical calculations, but also to support of DBs on substances and materials properties developed in Japan [18] and to integrate them with foreign information systems and to take advantage of artificial intelligence methods application for new substances design [19, 20].

6 Chinese Materials Genome Initiative

This five-year project started in China in 2016 at support of Ministry of Science and Technology [21]. Previously between 2014 and 2015 several MGI centers were organized: Shanghai Institute of Materials Genome (2014), Beijing Key Laboratory for Materials Genome (2015) and International Institute of MGI in Ningbo (2015), etc. The project goals are similar to the US MGI.

7 Large-Scale Infrastructure Projects for Materials Science Information Support Analysis

There should be noted several general trends in information support systems development in materials science areas:

- Integrated network of DBs on materials and substances properties development;
- development and broad application of computational methods;
- DBs with calculated information on materials development.

The analysis of goals and their achievement methods shows that the projects of the USA and China are most promising. In future, they could allow creation of full-fledged infrastructure for innovative activity information support in new materials development and deployment, having provided the science and the industry with reliable and complete data on substances and materials properties together with various tools (packages for quantum mechanical calculations, data mining, etc.) for substances parameters calculations. Japanese initiative is more limited than American one, because it is based on NIMS databases on materials and substances properties usage, and it also uses already known calculation methods experience (for example, VASP – widely known package for quantum mechanical calculations [22]). Investigations on artificial intelligence methods application were begun [19, 20]. Besides, the Japanese specialists are limited in a research field since they consider materials for electronics only (power supplies, magnetic, thermoelectric and spintronic materials) [17]. The EU projects at their initial stage seems to be the least promising. Orientation to the American DBs on substances properties and together with only quantum mechanical calculations significantly reduces these infrastructure projects potential and opportunities.

Nevertheless, it should be noted that to successfully implement of offered in the USA, the EU, China and Japan initiatives it is required, on the one hand, to get a breakthrough in materials properties calculation methods development, and, on the other hand, to achieve a progress in availability of databases on substances properties developed in recent years in different countries. The overview of available DBs on inorganic chemistry and materials science is given in the article [23] and in the IRIC (Information Resources on Inorganic Chemistry) information system [24]. In spite of the fact that hundreds of millions of dollars are spent for materials science information systems creation and support, their use is economically profitable since they allow reducing costs for new materials development considerably due to researches duplication reduction and reliable online information on substances properties providing to chemists and materials scientists. In turn, calculation methods give us a chance for substances parameters 'a priori' estimation, of prediction of substances, promising for industry applications, and for development of technology for materials production and processing. The consequence of these tasks solution is cost and time reduction for new materials development and deployment.

8   Integrated Information System on Inorganic Substances and Materials Properties Development Experience

Russian investigators undertake attempts to create their own materials infrastructure for different application areas. Premises for materials infrastructure project successful accomplishment in Russia are an experience in available from the Internet databases on inorganic substances and materials properties development and integration, together with expertise in methods and software for new substances and materials computer-aided design based on data mining technologies and, first of all, precedent based pattern recognition methods [23, 25, 26].

It should be noted that interest in data mining methods application to inorganic materials science is connected with objective difficulties arising at quantum mechanical calculations for yet not synthesized multicomponent inorganic substances, especially in a solid phase. For example, to calculate inorganic compound electronic structure by means of VASP package [22], it is necessary to know its crystal structure, i.e. it is necessary to produce and investigate this substance. Using pattern recognition methods for information analysis on already known substances from DBs it is possible to predict yet not synthesized substances and to estimate some of their properties, having only well-known components parameters (chemical elements or simple compounds). The special Information-Analytical System (IAS) (Fig. 1) was developed to solve this task in IMET RAS. The system includes the integrated databases system on inorganic substances and materials properties, the subsystems for regularities search in data and new substances prediction and their properties estimation, the knowledge base, predictions base and other subsystems [23, 26].

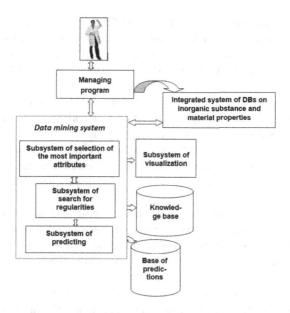

Fig. 1. Information-analytical system for inorganic compounds design structure.

8.1   The Integrated System of Databases on Inorganic Substances and Materials Properties

The integrated system of databases on inorganic substances and materials properties currently consolidates several information systems developed in IMET RAS [23, 26]: database on phase diagrams of semiconductor systems ("Diagram"), DB on substances with significant acousto-optical, electro-optical and nonlinear optical properties ("Crystal"), DB on inorganic substances forbidden zone width ("Bandgap"), DB on inorganic compounds properties ("Phases") and DB on chemical elements properties ("Elements"), and also "AtomWork" - DB on inorganic substances properties, developed in National Institute for Materials Science (NIMS, Japan), and "TKV" - DB on substances thermal constants developed in JIHT RAS and Lomonosov Moscow State University cooperation. This virtually integrated system allows providing access for user to wide variety of materials science information:

– "Phases" DB on inorganic compounds properties [27, 28] currently contains information on properties of more than 52 thousand ternary compounds (i.e. the compounds formed by three chemical elements) and more than 31 thousand quaternary compounds, obtained from more than 32 thousand publications. It includes summary for the most widespread inorganic compounds properties: crystal-chemical (type of crystal structure with indication of temperature and pressure above which the specified structure is formed, a crystal system, space group, number of formula units in unit cell, crystal lattice parameters) and thermophysical (melting type, melting and boiling points and compound dissociation temperature in solid or gas phases at an atmospheric pressure). In addition, the DB contains information on compounds superconducting properties. The "Phases" DB is formed by materials science experts on the basis of data analysis of periodicals, handbooks, monographs, reports, and abstract journals also (more than a half of sources are stored in PDF documents form). Currently "Phases" DB volume exceeds 25 GBytes, and it is available online to registered users from the Internet [28].

- "Elements" DB [23, 29] contains information on 90 most widespread chemical elements properties: thermophysical (melting and boiling points at 1 atmosphere, standard heat conductivity, molar heat capacity, an atomization enthalpy, entropy, etc.), dimensional (ionic, covalent, metal, pseudo-potential radii, atom volume, etc.), other physical properties (a magnetic susceptibility, electrical conductivity, hardness, density, etc.), position in the Periodic table of elements, etc. The DB is available online from the Internet [29].

- "Diagram" DB [30, 31] contains the information collected and estimated by highly skilled experts on phase P-T-x-diagrams of binary and ternary semiconductor systems and about physical and chemical properties of the phases which are formed in them. The DB volume exceeds 2 GBytes. The DB is available online to the registered users from the Internet [31].

- "Bandgap" DB [32, 33] contains information (more than 0.7 GBytes) on the forbidden zone width for more than 3 thousand inorganic substances. The DB is available online to users from the Internet [33].

- "Crystal" DB [34, 35] contains information on properties (piezoelectric (piezoelectric coefficients, elastic constants, etc.), nonlinear optical (nonlinear optical coefficients, Miller tensor components, etc.), crystal-chemical (crystal structure type, a crystal system, space group, number of formula units in unit cell, crystal lattice parameters), optical (refraction indices, transparency band, etc.), thermophysical (melting point, heat capacity, heat conductivity, etc.), etc.) for more than 140 acousto-optical, electro-optical and nonlinear optical substances collected and estimated by highly skilled experts in this subject domain. The DB volume exceeds 4 GBytes. It supports Russian and English languages for user interface and it's available online to the registered users from the Internet [35].

- "Inorganic Material Database – AtomWork" DB [36, 37] contains information on more than 82 thousand crystal structures, 55 thousand materials properties values and 15 thousand phase diagrams. The DB is available online to users from the Internet [37].

- "TKV" DB on substances thermal constants [38] contains information, available online from the Internet, on about 27 thousand substances formed by all chemical elements.

Taking into account current development trends in materials science databases, the complex integration approach that combines integration at data and user interfaces level is applied to materials database consolidation.

When integrating databases at user interfaces (actually, Web applications) level it's required to provide facilities for browsing information contained in other databases. This information should be relevant to the data on some chemical system currently being browsed by user. For example, user who browses information on Ga-As system from "Diagram" database should have an opportunity to get information for example on piezoelectric effect or nonlinear optical properties of GaAs substance contained in "Crystal" database. To achieve this type of behavior, it's required to provide search for relevant information contained in other databases of distributed system. Thus, it's necessary to have some active data center that should know what information is contained in every integrated database. Obviously, some data store should exist that somehow describes information on chemical entities, contained in integrated database resources. In this manner, metabase concept appears – a special database, that contains some reference information on integrated databases contents [39].

Metabase determines integrated system capabilities. Its structure should be flexible enough to represent metadata information on chemical entities of integrated DBs on inorganic materials properties. Taking into consideration the fact that chemical entities and their corresponding properties description is given at different detail level in different DBs, it's important to develop metabase structure that would be suitable for description of information residing in different DBs on inorganic substances properties. For example, some integrated DBs contain information on particular crystal modifications properties while others contain properties description at chemical system level. Thus, integrated DBs deal with different chemical entities. Relation between chemical entities can be described by means of chemical entities hierarchy in tree form (Fig. 2).

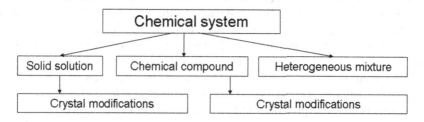

Fig. 2. Chemical entities hierarchy.

Having designated second level entities by the "substance" term, we get three-level chemical entities hierarchy: chemical system, chemical substance and crystal modification [39]. As far as information stored in DBs on inorganic substances properties can be considered at chemical system level, for simplicity in the paper this level will be taken from the top of the entities hierarchy.

The relevant information search problem can be formalized in terms of set-theoretic approach. Hence, metabase should contain information on integrated databases (D set), information on chemical substances and systems (S set) and information on their properties (P set). To describe correlation between elements of D, S and P sets let's define ternary relation called W on set $U = D \times S \times P$. Here U is a Cartesian product of D, S and P. Membership of a (d, s, p) triplet to the W relation, where $d \in D$, $s \in S$, $p \in P$. can be interpreted in the following way: "Information on property p of chemical system s is contained in integrated database d". Having defined three basic sets it can be seen that search for information relevant to s system can be localized to determination of R relationship, that is a subset of Cartesian product $S \times S$ (or in other words, $R \subseteq S^2$). Thus, it can be stated about every pair $(s_1, s_2) \in R$ that chemical system s_2 is relevant to the system s_1. So, to solve the task of relevant information search in integrated databases it's required to define the R relation. It is significant to note that R relation can be created or complemented by means of either of two variants. The first variant is via using predefined rules by a computer. The second one is that experts in chemistry and materials science can be engaged to solve this task.

The second variant is quite clear – experts can form relationship R manually following some multicriterion rules affected by their expert assessments. To consider one of the simplest ways to automatic R relation generation, it's required to implement a couple of rules, based on substances composition [39]:

1. For any chemical systems $s_1 \in S$, $s_2 \in S$ composed from chemical elements e_{ij}, $s_1 = \{e_{11}, e_{12}, ..., e_{1n}\}$, $s_2 = \{e_{21}, e_{22}, ..., e_{2m}\}$ it is true, that if $s_1 \subseteq s_2$ (i.e. all chemical elements of system s_1 are contained in system s_2), then $(s_1, s_2) \in R$. (1)
2. R relation is symmetric. In other words, for any $s_1 \in S$, $s_2 \in S$, it is true, that if $(s_1, s_2) \in R$, then $(s_2, s_1) \in R$ as well. (2)

These two rules allow determination of a set of chemical systems relevant to the given one. It should be noted that this automatic R relation generation variant is just one of the simplest and most obvious variants of such rules, and in fact more complex mechanisms can be used to get R relation. For example, browsing information on a particular property of a compound in one of integrated databases (in fact, it is information defined by (d_1, s_1, p_1) triplet), it is possible to consider (d_2, s_2, p_2) triplet to be relevant information. (d_2, s_2, p_2) triplet characterizes information on some other property of a system from another integrated database. In this case, more complex relevance relation arises: $R \ni (d_1, s_1, p_1) \times (d_2, s_2, p_2)$, where $d_1, d_2 \in D$, $s_1, s_2 \in S$, $p_1, p_2 \in P$. In fact, it's possible to define even a set of several R relations $(R_1, R_2, ..., R_n)$ by applying different rules. Thus, potentially it's possible to perform search for relevant information based on wide variety of R interpretations to provide required flexibility if needed.

To provide secure and transparent user transitions to relevant information DB Web-application it is planned to use special metabase gateways. To consider an example of integrated system functioning let's assume that in one of integrated system a user browses information on some particular chemical system. In other words, the user is in Web application of a particular information system. If it is necessary to get relevant information, the Web application will be capable to send a request to specially developed Web service that serves integrated system users. The request goal is to get information contained in integrated resources that is relevant to the currently browsed data. After the request the Web service sends a response to the Web application in a form of XML document. It describes what relevant information on chemical systems and properties is contained in integrated resources. It's well-known, that data in XML format are properly understood on all major platforms. That information can be output to user for example by means of a XSL-transformation in form of HTML document (XML + XSL = HTML) containing hyperlinks to special gateway [39]. The user could follow from one Web application to another to browse relevant information via this gateway only.

The gateway is a specialized Web application that runs on the metabase Web server. The gateway main purpose is to perform security-dispatching function in distributed system. According to the task stated it is responsible for user authentication and it also checks whether the user has required privileges to address the information requested. If that authentication is successful, i.e. the user is eligible to address the data, then the metabase security gateway will perform redirection to a specialized entry point of desired Web application adding some additional information to create proper security context in target Web application and supplying it with digital signature. It should be stated that the entry point is a specialized page in target Web application that is to perform service functions for integrated system users. At this page target Web application checks digital signature of the metabase security gateway and if everything is fine the page will create special security context for user with given access privileges within target Web application. Finally, the user is automatically redirected to the page with the information required. In spite of redirection process apparent complexity, user transition from one Web application to another is absolutely transparent [39]. Thus, end user can even not note that some complex processing has been done to perform redirection. So, it is an illusion created that having clicked on a hyperlink the user is transferred from one information system to another directly.

8.2 Inorganic Compounds Computer-Aided Design System

The inorganic compounds computer-aided design system background is formed by precedent-based pattern recognition algorithms and programs which are collected within the multipurpose "Recognition" system, developed by Dorodnicyn Computing Centre of RAS [40] and combining in addition to widely known methods of linear machine, linear Fischer discriminant, k-nearest neighbors, support vector machine, neural networks and genetic algorithms, also a number of unique algorithms developed in Dorodnicyn Computing Centre of RAS: the recognition algorithms based on estimates calculations, deadlock tests voting algorithms, logical patterns voting algorithms, statistical weighed voting algorithms, etc. The inorganic compounds computer-aided design system also includes ConFor system, developed at Institute of Cybernetics of National Academy of Sciences of Ukraine [41]. The ConFor is a software for machine learning based on so-called growing pyramidal networks which are special data structures in computer memory that form subject domains concepts.

To select the most important components properties IAS includes several programs based on various algorithms [42–44]. The developed system deployment makes it possible to predict new inorganic compounds and estimate various properties of those without experimental synthesis [25, 26, 32].

9 The Project of Infrastructure for Providing Russian Specialists with Data in the Field of Inorganic Chemistry and Materials Science

IAS is, some kind of, pilot project for creation of information infrastructure for inorganic materials science. According to this project the most known Russian DBs in this area are virtually integrated, and also integration of Russian systems with foreign information systems was begun. The majority of Russian DBs contain references to full publications texts from which information of DBs was extracted. The compound computer-aided design subsystem allows searching for regularities in DB information and applying them to yet not synthesized substances prediction and their properties estimation. It should be noted that at a prediction phase only data on compounds components properties (chemical elements or simple compounds) are used. Obtained predictions are stored in special predictions base that expands traditional databases functionality (user obtains not only well-known experimental data, but also predictions for yet not synthesized compounds and some of their properties estimation).

When developing the Russian infrastructure project for specialists in the field of inorganic materials science information support it is necessary to consider all possible user queries variety. It is natural that academic scientist's queries could significantly differ from queries of design engineers or materials producers. However the general information infrastructure project should necessarily include virtually integrated system of Russian and foreign databases on inorganic substances and materials properties, production and processing technologies, materials producers and consumers, etc., a complex of packages for materials modeling and calculation which are widely used in most cases by academic scientists, and virtually integrated databases system with already calculated values to simplify new materials industry adoption (Fig. 3). It is necessary to emphasize that technologies of processing, storage and search of necessary data require development and usage of the most modern software and powerful data-processing centers (DPC) creation.

Materials DB system should consolidate factual DB on inorganic substances and materials that are the most important for Russian users (Russian DBs: IMET RAS, JIHT RAS, MSU, etc. and foreign DBs: NIMS [18], NIST [11], STN [45], Springer Materials [46], etc.), the leading publishing corporations documentary DBs (Science, Elsevier, Springer, Wiley, American Chemical Society, American Institute of Physics, Science, etc.), and also databases with yet not published information (All-Russian Institute of Scientific and Technical Information, Center of Information Technologies and Systems, etc.), patent databases (Rospatent, Questel, USPTO, etc.), databases on inorganic materials producers and consumers, etc. It is necessary to allocate funds for annual licenses prolongation for foreign materials databases usage and to organize the uniform portal with free access for Russian users (currently such databases are available to limited organizations only). It is necessary to support in every possible way transfer to an electronic form of paper collections of popular Russian scientific journals, which are the most in the world (in addition it will undoubtedly promote them and increase their authority and impact-factors).

To equip research organizations with intellectual calculation systems it's required, first of all, to start with students and graduate students training for use of the most known quantum mechanical, thermodynamic, statistical, etc. calculations packages. It is necessary to develop Russian databases with calculated values and to integrate them with foreign information systems, which are available in the Internet currently (for example, [15]) that will allow partially solution of a problem of qualified calculations on substances properties. Experiment planning should include calculations usage at initial research stage that will allow reduction of time and costs for new materials search and development.

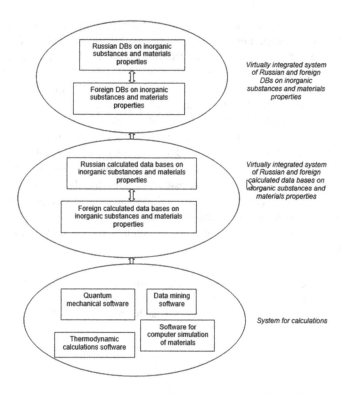

Fig. 3. The scheme of infrastructure for providing Russian specialists with data in the field of inorganic chemistry and materials science.

User queries analysis subsystem, especially, for specialists in applied areas should become an important component of the developed infrastructure project. It will allow discovery of special interest materials groups to which research and investigation should be aimed to. The statistics of absence in user requested values of a particular substance parameter can become a signal to initiate an additionally required property investigation.

10   Conclusion

Transition of the Russian economy to an innovative way of development and increase in product competitiveness is defined in many respects by materials quality, novelty and functionality. At the present technologies development stage, new materials search, research and implementation requires mature infrastructure creation including the academic organizations with their new substances theoretical and pilot investigation potential, the organizations conducting applied researches on development and deployment of new materials and their production and processing technologies, the shareable scientific equipment centers with expensive equipment complexes including mega-science objects, etc. In recent years in the developed countries several projects were initiated (MGI, MDF, NoMaD, etc.). These projects are of strategic importance for achieving technological excellence by creating infrastructure for new materials with predefined functional properties set development and deployment acceleration. Special attention in these projects is paid to information support infrastructure. The Russian answer to these strategic initiatives of the USA, the EU, Japan, China in materials information infrastructure can be creation of the federal information center providing specialists with information on substances and materials properties, corresponding production technologies, and calculated properties values also, patent information, etc. In connection with considered subject domain peculiarities the distributed integrated network of Russian and foreign databases and knowledge bases on substances and materials should form a basis of such shared information center. Federal information center creation that integrates materials science information resources will stimulate faster new materials search, development and deployment. In a combination with considerable cost reduction due to researches duplication elimination it also will provide chemists and materials scientists with operational and reliable experimental and calculated information on substances and materials.

Acknowledgements. The authors thank A. V. Stolyarenko, V. V. Ryazanov, O. V. Sen'ko, A. A. Dokukin for their help in an information-analytical system development. Work is partially supported by the Russian Foundation for Basic Research, projects 16-07-01028, 17-0701362 and 15-07-00980.

References

1. Materials Genome Initiative: "Strategic Plan. National Science and Technology Council. Committee on Technology", Subcommittee on the Materials Genome Initiative. https://www.whitehouse.gov/sites/default/files/microsites/ostp/NSTC/mgi_strategic_plan__dec_2014.pdf
2. Kiselyova, N.N., Dudarev, V.A.: Inorganic chemistry and materials science data infrastructure for specialists. In: Selected Papers of the XVIII International Conference on Data Analytics and Management in Data Intensive Domains (DAMDID/RCDL 2016), vol. 1752, pp. 121–128. CEUR Workshop Proceedings (2016)
3. Kalinichenko, L.A., Volnova, A.A., Gordov, E.P., Kiselyova, N.N., et al.: Data access challenges for data intensive research in Russia. Informatika i ee Primeneniya – Inf. Appl. **10**(1), 3–23 (2016)
4. Materials Genome Initiative for Global Competitiveness. http://www.whitehouse.gov/sites/default/files/microsites/ostp/materials_genome_initiative-final.pdf
5. Materials Genome Initiative. https://www.mgi.gov/partners
6. Curtarolo, S., Setyawan, W., Wang, S., et al.: AFLOWLIB.ORG: a distributed materials properties repository from high-throughput ab initio calculations. Comput. Mater. Sci. **58**, 227–235 (2012)
7. Taylor, R.H., Rose, F., Toher, C., et al.: RESTful API for exchanging materials data in the AFLOWLIB.org consortium. Comput. Mater. Sci. **93**, 178–192 (2014)
8. University of Chicago: Microscopic animals inspire innovative glass research. http://www.uchicago.edu/features/microscopic_animals_inspire_innovative_glass_research/
9. The First Five Years of the Materials Genome Initiative: Accomplishments and Technical Highlights (2016). https://mgi.nist.gov/sites/default/files/uploads/mgi-accomplishments-at-5-years-august-2016.pdf

10. National Data Service: The Materials Data Facility. https://www.materialsdatafacility.org
11. NIST Data Gateway. NIST Online Databases. http://srdata.nist.gov/gateway/gateway?dblist=0
12. Saal, J.E., Kirklin, S., Aykol, M., et al.: Materials design and discovery with high-throughput density functional theory: the Open Quantum Materials Database (OQMD). JOM **65**(11), 1501–1509 (2013)
13. The Novel Materials Discovery (NOMAD) Laboratory. http://nomad-lab.eu/
14. The Novel Materials Discovery (NOMAD) Laboratory. EINFRA-5-2015 - Centres of Excellence for computing applications. http://cordis.europa.eu/project/rcn/198339_en.html
15. The NoMaD Repository. http://nomad-repository.eu/cms/
16. Materials design at the eXascale. http://cordis.europa.eu/project/rcn/198340_en.html
17. Center for Materials Research by Information Integration. http://www.nims.go.jp/eng/research/MII-I/index.html
18. NIMS Materials Database (MatNavi). http://mits.nims.go.jp/index_en.html
19. Lee, J., Seko, A., Shitara, K., Tanaka, I.: Prediction model of band-gap for AX binary compounds by combination of density functional theory calculations and machine learning techniques. Phys. Rev. B **93**(11), 115104 (2016)
20. Toyoura, K., Hirano, D., Seko, A., et al.: Machine-learning-based selective sampling procedure for identifying the low-energy region in a potential energy surface: a case study on proton conduction in oxides. Phys. Rev. B **93**(5), 054112 (2016)
21. Lu, X.-G.: Remarks on the recent progress of Materials Genome Initiative. Sci. Bull. **60**(22), 1966–1968 (2015)
22. The Vienna Ab initio Simulation Package (VASP). https://www.vasp.at/
23. Kiselyova, N.N., Dudarev, V.A., Zemskov, V.S.: Computer information resources in inorganic chemistry and materials science. Russ. Chem. Rev. **79**(2), 145–166 (2010)
24. IRIC DB (Information Resources on Inorganic Chemistry). http://iric.imet-db.ru/
25. Kiselyova, N.N.: Computer design of inorganic compounds. Application of databases and artificial intelligence. Nauka, Moscow (2005)
26. Kiselyova, N.N., Dudarev, V.A., Stolyarenko, A.V.: Integrated system of databases on the properties of inorganic substances and materials. High Temp. **54**(2), 215–222 (2016)
27. Kiselyova, N., Murat, D., Stolyarenko, A., et al.: Phases database on properties of ternary inorganic compounds on the Internet. Inf. Res. Russ. **4**, 21–23 (2006)
28. "Phases" DB. http://www.phases.imet-db.ru
29. "Elements" DB. http://phases.imet-db.ru/elements
30. Khristoforov, Y.I., Khorbenko, V.V., Kiselyova, N.N., et al.: The database on semiconductor systems phase diagrams with Internet access. Izv. Vyssh. Uchebn. Zaved. Mater. Electron. Tech. **4**, 50–55 (2001)
31. "Diagram" DB. http://diag.imet-db.ru
32. Kiselyova, N.N., Dudarev, V.A., Korzhuyev, M.A.: Database on the bandgap of inorganic substances and materials. Inorg. Mater. Appl. Res. **7**(1), 34–39 (2016)
33. "Bandgap" DB. http://www.bg.imet-db.ru
34. Kiselyova, N.N., Prokoshev, I.V., Dudarev, V.A., et al.: Internet-accessible electronic materials database system. Inorg. Mater. **42**(3), 321–325 (2004)
35. "Crystal" DB. http://crystal.imet-db.ru
36. Xu, Y., Yamazaki, M., Villars, P.: Inorganic materials database for exploring the nature of material. Jpn. J. Appl. Phys. **50**(11), 11RH02/1-5 (2011)
37. "AtomWork" DB. http://crystdb.nims.go.jp/index_en.html
38. "TKV" DB. http://www.chem.msu.su/cgi-bin/tkv.pl?show=welcome.html/welcome.html

39. Dudarev, V.A.: Information systems on inorganic chemistry and materials science integration. Krasand, Moscow. 320 p. (2016)
40. Zhuravlev, Y.I., Ryazanov, V.V., Senko, O.V.: Recognition. Mathematical methods. Program system. Practical applications. FAZIS, Moscow. 176 p. (2006)
41. Gladun, V.P.: Processes of forming of new knowledge. SD "Pedagog-6", Sofia. 186 p. (1995)
42. Senko, O.V.: An optimal ensemble of predictors in convex correcting procedures. Pattern Recogn. Image Anal. **19**(3), 465–468 (2009)
43. Yuan, G.-X., Ho, C.-H., Lin, C.-J.: An improved GLMNET for L1-regularized logistic regression. J. Mach. Learn. Res. **13**, 1999–2030 (2012)
44. Yang, Y., Zou, H.: A coordinate majorization descent algorithm for L1 penalized learning. J. Stat. Comput. Simul. **84**(1), 1–12 (2014)
45. STN website. http://www.stn-international.de/
46. Springer Materials. http://materials.springer.com/

Visual Analytics of Multidimensional Dynamic Data with a Financial Case Study

Dmitry D. Popov[1(✉)], Igal E. Milman[1], Victor V. Pilyugin[1],
and Alexander A. Pasko[2]

[1] National Research Nuclear University MEPhI, Moscow, Russia
{DDPopov,VVPilyugin}@mephi.ru, igalush@gmail.com
[2] The National Centre for Computer Animation,
Bournemouth University, Poole, UK
apasko@bournemouth.ac.uk

Abstract. This work deals with a problem of analysis of time variant objects. Each object is characterized by a set of numerical parameters. The visualization method is used to conduct the analysis. Insights of interest for the analyst about the considered objects are obtained in several steps. At the first step, a geometric interpretation of the initial data is introduced. Then, the introduced geometrical model undergoes several transformations. These transformations correspond to solving the first problem of the visualization method, in particular, obtaining visual representations of data. The next step for the analyst is to analyze the generated visual images and to interpret the results in terms of the considered objects. We propose an algorithm for the problem solution. Developed interactive visualization software is described, which implements the proposed algorithm. We demonstrate how with this software the user can obtain insights regarding the creation and disappearance of object clusters and bunches and find invariants in the initial data changes.

Keywords: Visual analytics · Scientific visualization · Data visualization · Multidimensional data · Financial data

1 Introduction

Visual analytics is the one of different forms of data analysis. James Thomas wrote in [3]: "Visual analytics is the science of analytical reasoning facilitated by interactive visual interfaces." This work turned researchers' attention to this new interdisciplinary science area.

Solving problems using the method of visualization is one of visual analytics forms. Theoretical generalizations of this method for the case of scientific data analysis were presented in [7]. In general, it assumes a sequence of two tasks, which should be solved iteratively:

1. Data visualization (obtaining images);
2. Image analysis with generation of insights about the initial data values.

© Springer International Publishing AG 2017
L. Kalinichenko et al. (Eds.): DAMDID/RCDL 2016, CCIS 706, pp. 237–247, 2017.
DOI: 10.1007/978-3-319-57135-5_17

In this work, some theoretical and practical issues are considered from the multi-dimensional dynamic data analysis point of view. We define multidimensional dynamic data as numerical data presented in the tabular form. Each table consists of data related to the state of a dynamic object at some time moment.

This article is the elaboration of [9]. In Sect. 2, we present a general approach to data analysis and specify the problem statement of the research. The description of transformations in Sect. 4 has been changed, besides, the idea of discretization of processes has been defined more precisely. Section 5 contains the updated description of steps of the visualization method. In Sect. 6, we, in more details, provide the algorithm to solve the assigned task. Explanations of the first example in Sect. 7 has been added. In the conclusion, we altered the future development of the research.

2   Approaches to Multidimensional Data Analysis

During the last thirty years, a lot of papers and books on visualization approaches have appeared. Many works about nanostructure dynamics like [1, 2] may be found among papers describing multidimensional data analysis. Characteristics of spatial-temporal structures are analyzed in [5]. Three-dimensional representations can be constructed for objects considered in these papers. The problem of visualization for data having more than four dimensions is mentioned in [5].

An approach to visualization of data on patient health is proposed in [8]. One object described by more than three parameters changing in time is analyzed. Our method described in this article allows for analysis of a time variant set of multidimensional objects with higher dimensional objects being of more practical interest.

Two or three data components from [4] have a spatial nature. Software used to deal with such problems is called a Geo-Information System (GIS). Each object of interest is presented with a geometric image on a 2D or 3D map in such systems. For example, a geometric image can be a circle, a square, a ball, a cube or a line showing a trajectory of object motion. Non-spatial data components are shown in separate windows of graphical user interface. However, they can be indicated on the map using optical parameters (color, transparency) or sizes and shapes of geometric images.

Multidimensional data analysis methods and software mentioned above differ by initial data models and approaches to visualization. These differences stem from the various tasks of data analysis. In this paper, data are explored for detection of similar objects in the set. Time variations of similar objects and their data properties are researched too.

3   Problem Statement

A set of m objects is given. Each object is characterized by n parameters with k tables to the following form initially assigned:

Each table consists of values of object parameters at the time t_j, $t_1 < t_2 < \ldots < t_k$. x_{il}^j is a value of l-th parameter of i-th object at the time $t_j (l = \overline{1,n}, i = \overline{1,m}, j = \overline{1,k})$ (Table 1).

Table 1. j-th table with data about object parameter values

	Parameter 1	Parameter 2	...	Parameter n
Object 1	x_{11}^j	x_{12}^j	...	x_{1n}^j
Object 2	x_{21}^j	x_{22}^j	...	x_{2n}^j
...	...	...	...	...
Object m	x_{m2}^j	x_{m2}^j	...	x_{mn}^j

3.1   Geometrization of Data

To find the problem solution we need to introduce its geometrical interpretation. Initial data are presented as points of multidimensional Euclidian space E_n with a metric $\rho(x, y)$, where:

- Table row is a point of multidimensional Euclidian space.
- Table column contains values of one coordinate in the Euclidian space.
- Distance in the Euclidian space is interpreted as a disparity measure of objects.

Distance in the space helps us separate out subsets of points: clusters and bunches. Definitions of these subsets are declared in [6]. Values of point coordinates are assigned at times $t_1, t_2, \ldots, t_k$.

Let us introduce a definition of a geometrical process. The geometrical process is a set of space points with time dependent coordinates. Thus, discrete geometrical process $P = P(t_j)$ is initially given with $P(t_j)$ defined on a set of discrete time values $T = \{t_1, t_2, \ldots, t_k\}$.

$$P(t_j) = \{p_1(t_j), p_2(t_j), \ldots, p_m(t_j)\},$$

where $p_i(t_j)$ is an n-dimensional point:

$$p_i(t_j) = \left(p_i^1(t_j), p_i^2(t_j), \ldots, p_i^n(t_j) \right),$$

with $p_i^l(t_j) = x_{il}^j$ – value of l-th coordinate of i-th point. The distance is defined for each pair of points as:

$$\rho\left(p_{i_1}(t_j), p_{i_2}(t_j) \right) \geq 0, i_1, i_2 = \overline{1, m}, i_1 \neq i_2.$$

The type of chosen metric depends on the nature of initial data. Many research papers describe the situation when metric was empirically chosen while data were investigated to better suit the analysis task, for example, clustering. In this work, the developed software, which will be described later, has options of working with Euclidean distance, Minkowski distance, Chebyshev distance, in particular, and Mahalanobis distance.

4 Transformation of Geometrical Processes

A discrete geometrical process transformation is necessary for the problem solution. The process $P(t_j)$ is a time sequenced discrete set of descriptions of the states of objects under investigation. Meanwhile, objects themselves exist and change continuously. Therefore, we need a tool to transform the discrete process into a continuous one. This tool is an interpolation.

4.1 Interpolation

The interpolation consists in the search of a function F from the given function class such that:

$$F(t_j) = P(t_j),$$

with $t_j \in T$. The search of a function F means the search of functions f_i^l from the given function class such that:

$$f_i^l(t_j) = p_i^l(t_j),$$

since

$$P(t_j) = \{p_1(t_j), p_2(t_j), \ldots, p_m(t_j)\},$$

$$p_i(t_j) = (p_i^1(t_j), p_i^2(t_j), \ldots, p_i^n(t_j)).$$

In the case of continuous functions f_i^l, the initial discrete geometrical process interpolation results in a continuous geometrical process $P(t)$. If the function f_i^l is an algebraic binomial, then for any value $\in [t_h, t_{h+1}], 1 \le h \le k - 1$, f_i^l is given by the following equation:

$$f_i^l(t) = p_i^l(t_h) + \frac{p_i^l(t_{h+1}) - p_i^l(t_h)}{t_{h+1} - t_h}(t - t_h).$$

The interpolation using an algebraic binomial (piecewise-linear interpolation) is the simplest type of interpolations. Functions interpolated by the piecewise-linear interpolation method are continuous functions. The piecewise-linear interpolation is used with missing priori information about time dependency of initial data. This type of interpolation does not require heavy computations and well describes the observed dependence with small time intervals.

4.2   Discretization of Continuous Geometrical Processes

$P(t_0)$ is denoted a time sample of the continuous geometrical process $P(t)$ at time t_0 from the $P(t)$ domain. A discrete geometrical process can be constructed from $P(t)$ using time samples at times $\tau_1, \tau_2, \ldots, \tau_{k'}$:

$$P'(\tau_h) := \{P(\tau_h)|\tau_h \in \{\tau_1, \tau_2, \ldots, \tau_{k'}\}\}.$$

5   The Visualization Method for the Problem Solution

The visualization method is a sequential solution of two tasks, which is shown in Fig. 1.

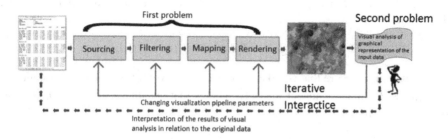

Fig. 1. Visualization method

One of the most important concepts of the visualization method is a spatial scene, which is used for solving the first task. The spatial scene is a set of spatial objects with their geometric and optical parameters. The descriptions of geometric parameters of spatial objects such as shape, size and position constitute a geometric model of the scene. Optical parameters of spatial objects such as color or transparency are named an optical model of the scene.

The first task solution implies that initial data have been mapped to the spatial scene. Success of analysis depends on correctness of the mapping. Geometric models which would support data mining should be chosen at this stage. On the one hand, these models must be chosen according to the goals of analysis, for example, shape or mutual position analysis. Additionally, they must enable interpretation of visual analysis results in respect to initial data.

Changing visualization attributes and getting images or animations are carried out until a decision about initial data has been reached. Let us take a closer look at the data visualization process (Fig. 1).

5.1   Sourcing

Values of multidimensional point coordinates at different times $P(t_j)$ are received at this stage.

5.2   Filtering

Initial data are preprocessed at this stage. $P(t_j)$ is interpolated linearly for the task solving using the method described in Sect. 4.1. Filtering gives continuous dependency $P(t)$.

5.3   Mapping

Initial data are mapped to a dynamic spatial scene

$$S(t) = \;<G(t), O(t)>$$

where $G(t)$ denotes the geometric model of the scene and $O(t)$ denotes the optical model of the scene. For any value of t in the interval $[t_1, t_k]$, $S(t)$ corresponds to $P(t)$. The geometric model:

$$G(t) = \{Sph_1(t), \ldots, Sph_m(t), Cyl_1(t), \ldots, Cyl_e(t)\}$$

where $Sph_i(t)$ is a sphere that is linked to point $p_i(t)$; $Cyl_j(t)$ is a cylinder that connects two spheres. Cylinders connect the only spheres that are linked to $p_{i_1}(t_j), p_{i_2}(t_j)$ satisfying the condition:

$$\rho\left(p_{i_1}(t_j), p_{i_2}(t_j)\right) \leq d$$

where ρ is a distance between $p_{i_1}(t_j), p_{i_2}(t_j)$, d is a constant that is defined by the analyst. The optical model:

$$O(t) = \{SphC_1(t), \ldots, SphC_m(t), CylC_1(t), \ldots, CylC_e(t)\},$$

where $SphC_1(t) = \ldots = SphC_m(t) = SphC$ are colors of spheres, $SphC$ is defined by the analyst during the problem solution process, $CylC_1(t), \ldots, CylC_e(t)$ are colors of cylinders. The cylinder's color value in RGB in our implementation is calculated form the equations:

$$R(t) = 255\left(1 - \frac{\rho(p_{1_1}(t), p_{1_2}(t))}{d}\right),$$

$$G(t) = 150\frac{\rho(p_{1_1}(t), p_{1_2}(t))}{d},$$

$$B(t) = 255\frac{\rho(p_{1_1}(t), p_{1_2}(t))}{d}.$$

5.4 Rendering

The concept of rendering is defined in computer graphics. Rendering results in a projection image I of the scene S. The spatial scene correlates with time t and it means that projection images correlate with time too:

$$I(t) = I(S(t), A(t)),$$

where $A(t)$ are rendering attributes: camera, light sources, size of obtained image and others.

6 Description of the Problem Solution Algorithm

The algorithm of the problem solution consists of the following steps:

1. Input of initial dynamic data.
2. Interpolation and defining a continuous geometrical process.
3. Assigning of d and constant parameters of the scene: a radius of spheres, a radius of cylinders, a subspace for projection, color of spheres.
4. Constructing a dynamic spatial scene using a continuous geometrical process and the constant geometric parameters.
5. Discretization of a dynamic scene.
6. Assigning camera and other rendering attributes.
7. Image sequence acquisition using the scene at the time moments of the time set.
8. Saving of image sequence in a gallery or generating an animation.

If an analyst has not made a judgment using images or animations, then s/he can be back to the steps three and six.

The algorithm is shown in Fig. 2. Note that, static multidimensional data analysis is an important particular case in practice. There is no difficulty for understanding that analysis of a static multidimensional data is one kind of analysis of dynamic multidimensional data at a fixed time value, for example, in the case of $T = \{t_1\}$.

The algorithm implies the analyst-computer interaction, where responsibilities are distributed as follows:

- in the course of the distance analysis, computers are used to calculate distance values, to generate projections and plots of time-dependencies of point coordinates;
- analysts visually perceive projection images on the display, analyze mutual location of multidimensional points, reveal subsets and define visualization parameters.

The presented algorithm is implemented in the interactive software tool. Based on Autodesk 3ds Max, the tool employs the internal interpreted language MAXscript. A C# library written in Visual Studio 2013 was also employed. The software tool enables analysts to perform the following:

- Input of initial data (a description of the discrete geometrical process).
- Initial data are entered only once at a work-start time.
- Apply the piecewise-linear interpolation of the discrete geometrical process.

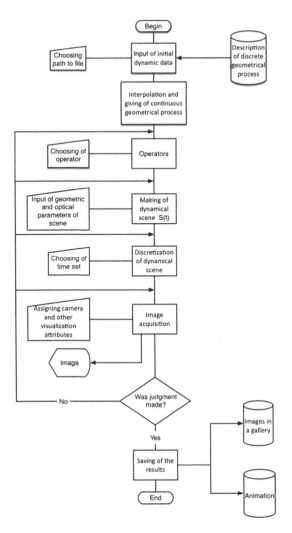

Fig. 2. Algorithm of the problem solution

- Construct a dynamic scene using the continuous geometrical process and obtain an animated visualization of it. Graphical projections are made in a standard 3ds Max window by means of a standard renderer.
- Retrieve point details, which can be viewed in a table with rows representing all displayed points and columns representing their coordinates. In the table, the rows have the same background color as corresponding points to let users easily identify a subset to which a particular point belongs. Table data can be generated for any time step under consideration.
- Measure the pairwise distance (in the initial n-dimensional space).
- Modify scene and rendering parameters. The scene parameters include radius of spheres correlating with initial points, their color, a cylinder radius and the 3D

projection space. The parameters are supposed to be defined at the starting point of the work and can be modified at any stage of the process.

- Manipulate the space scene with the use of standard 3ds Max tools. Among available tools are, for example:

  (a) Affine transformations of the scene;
  (b) Application of image filters;
  (c) Changes in optic characteristics of the scene.

- Merge a point into a subset. Each subset is mapped to a specific color allowing the user to discover the given subset in subsequent operations.
- Build and save planar graphical projections of multidimensional points. In the remote point analysis, it is important to define what coordinates mostly contribute to the distance – whether it is all of them or just a few with high dissimilarity. To clarify this, we suggest building graphical projections of the initial subset (p_i^l, p_i^l) and view all such projections by changing the l value.
- Obtain animation.
- Create plots of coordinate time-dependencies. In the case of a big dimension, projections of points can be very close to each other. Modification of spheres and cylinders radii is used to recognize close spheres. The decreasing radius allows the user to better analyze points. Moreover, if different points are projected to the same point of the three-dimensional space, then the software shifts centers of their spheres and paints them in different from $SphC$ color.

7   Financial Case Study of Software in Use

Analysis of 81 lending agencies was made with using the developed software. Data about the agencies consists of monthly reports of 9 activities of these agencies for 13 months of 2013 and 2014 years. Data of first reporting month are assigned as data at $t = 1$. Data of last reporting month are assigned as data at $t = 13$. Bunches and clusters

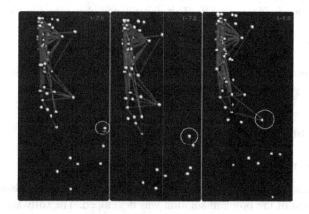

Fig. 3. Frames of generated animation

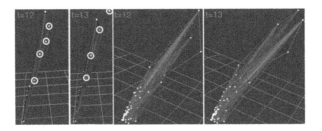

Fig. 4. Projection images of the spatial scene

membership of lending agencies in their geometrical interpretation were explored. Judgments about such clustering allows for reasoning on similarities between some lending agencies.

A series of images in Fig. 3 allow the analyst to make the following judgment. During the interval $t \in [7, 8]$ the remote point, which is marked by the circle, has joined the bunch.

As noted in [6], the described example is of relevance in the lending agencies audit. An analyst knows "good" agencies before the audit happens or s/he detects them in the audit process. An analyst can consider a cluster as "good" agencies. Further changes, which are happening with this subset, are analyzed. For example, points can be detected, which are joining the cluster or leaving it. Agencies related to these points are included in the special attention list of the analyst. An observation made by an analyst may result in a more detailed study of the organization. Changes within the bunch or cluster are important for the analyst too. Points with the rapidly changing distance in relation to other points during some time period are also of interest to the analyst. In other words, there are points which tend to approach the cluster boundary (limit) or points which tend to get closer to other point positions. These agencies also need keen attention. The analyst can detect these points watching colors of cylinders between spheres.

One of obtained scenes exposes the following law. Most of the spheres with the exception of marked by white circles in Fig. 4 lie on some plane. It allows us to extract the invariant dependency between three financial indicators: X – overdue debt in credit portfolio, Y – bounds, Z – net assets. This dependency obeys the equation of the plane:

$$2,231X - 72,672Y + 20,3624Z - 2,58513 = 0$$

The agencies related to the points marked with spheres do not conform to the law. The relation may be observed for the intervals $t \in [1, 9] \cup [12, 13]$.

8   Conclusion

Thus, in this work the following results were presented:

- Mathematical models of initial data and models of the operations were introduced.
- The algorithm of the problem solution was developed. The software tool based on the proposed algorithm was implemented.

- An analysis of multidimensional objects similarity was demonstrated with the financial case study of lending agencies.
- An opportunity of searching for an invariant in geometrical processes with using the software tool was shown.

An invariant is a property of explored data which always remains unchanged or which remains unchanged over a period of time. In the above example, the relation of three financial indicators remained the same with the dynamic changes. In the future development, new software functionality such as different interpolation procedures, user's scene modification tools and others can be added. For example, the scene modification can allow for construction of planes, spheres and other auxiliary geometric primitives. In this paper, classical multidimensional data analysis methods are not involved. However, the authors are studying approaches such as an intelligent analysis and data mining to further elevate the software system to a "numerical-visual" level.

References

1. Grottel, S., Reina, G., Vrabec, J., Ertl, T.: Visual verification and analysis of cluster detection for molecular dynamics. IEEE Trans. Visual. Comput. Graphics **13**(6), 1624–1631 (2007)
2. Sourina, O., Korolev, N.: Visual mining and spatio-temporal querying in molecular dynamics. J. Comput. Theor. Nanosci. **2**, 1–7 (2005)
3. Thomas, J., Cook, K.: Illuminating the Path: Research and Development Agenda for Visual Analytics. IEEE Press, New York (2005)
4. Wallner, G., Kriglstein, S.: PLATO: a visual analytics system for gameplay data. Comput. Graph. **38**, 341–356 (2014)
5. Bondarev, A.E.: Analysis of unsteady space-time structures using the optimization problem solution and visualization methods. Sci. Visual. **3**(2), 1–11 (2011). (in Russian)
6. Milman, I.E., Pakhomov, A.P., Pilyugin, V.V., Pisarchik, E.E., Stepanov, A.A., Beketnova, Y.M., Denisenko, A.S., Fomin, Y.A.: Data analysis of credit organizations by means of interactive visual analysis of multidimensional data. Sci. Visual. **7**(1), 45–64 (2015)
7. Pilyugin, V., Malikova, E., Adzhiev, V., Pasko, A.: Some theoretical issues of scientific visualization as a method of data analysis. In: Gavrilova, M.L., Tan, C.J.K., Konushin, A. (eds.) Transactions on Computational Science XIX. LNCS, vol. 7870, pp. 131–142. Springer, Heidelberg (2013). doi:10.1007/978-3-642-39759-2_10
8. Hachumov, V.M., Vinogradov, A.N.: Development of new methods for continuous identification and prognosis of the state of dynamic objects on the basis of intellectual data analysis. In: Proceedings of the 13th Russian Conference on Mathematical Methods of Image Recognition, Saint-Petersburg region, Zelenogorsk, pp. 548–550 (2007). (in Russian)
9. Popov, D.D., Milman, I.E., Pilyugin, V.V., Pasko, A.: Visual analytics of multidimensional dynamic data. In: Selected Papers of the 18th International Conference on Data Analytics & Management in Data Intensive Domains (DAMDID/RCDL 2016), Ershovo, Moscow Region, Russia, CEUR Workshop Proceedings, vol. 1752, pp. 51–57 (2016)

Metadata for Experiments in Nanoscience Foundries

Vasily Bunakov[1(✉)], Tom Griffin[1], Brian Matthews[1], and Stefano Cozzini[2]

[1] Science and Technology Facilities Council, Harwell, Oxfordshire, UK
{vasily.bunakov,tom.griffin,brian.matthews}@stfc.ac.uk
[2] CNR, Istituto Officina dei Materiali, Trieste, Italy
cozzini@iom.cnr.it

Abstract. Metadata is a key aspect of data management. This paper describes the work of NFFA-EUROPE project on the design of a metadata standard for nanoscience, with a focus on data lifecycle and the needs of data practitioners who manage data resulted from nanoscience experiments. The methodology and the resulting high-level metadata model are presented. The paper explains and illustrates the principles of metadata design for data-intensive research. This is value to data management practitioners in all branches of research and technology that imply a so-called "visitor science" model where multiple researchers apply for a share of a certain resource on large facilities (instruments).

Keywords: Metadata · Nanostructures foundries

1 Introduction

The Nanostructures Foundries and Fine Analysis (NFFA-EUROPE) project (www.nffa.eu) brings together European nanoscience research laboratories that aim to provide researchers with seamless access to equipment and computation. This will offer a single entry point for research proposals, and a common platform to support the access and integration of the resulting experimental data. Both physical and computational experiments are in scope, with a vision that they complement each other and can be mixed in the same identifiable piece of research.

Metadata design is a part of a joint research activity within NFFA-EUROPE that takes empirical input from the project participants, and also takes into account state-of-the art standards and practices. Metadata design is an incremental effort of the project; this work presents the first stage resulting in a high-level metadata model that is agnostic to the actual data management situation in participating organizations yet is able to capture significant features of physical and computational nanoscience experiments.

Compared to the well-known metadata recommendation for nanoscience developed by CODATA-VAMAS Working Group on the Description of Nanomaterials [7], which is heavily focused on nano-samples description, the metadata model we are developing in NFFA-EUROPE is intended to well reflect the lifecycle of data collected in nanoscience experiments (both physical and computational), and then archived for the purposes of further data discovery and data sharing. This is why this model makes the

© Springer International Publishing AG 2017
L. Kalinichenko et al. (Eds.): DAMDID/RCDL 2016, CCIS 706, pp. 248–262, 2017.
DOI: 10.1007/978-3-319-57135-5_18

most sense for data practitioners in nanoscience and for research users who want to discover and explore the context of data assets resulted from nanoscience experiments.

This work adds to the earlier published effort of metadata design for nanoscience [13]. It expands on the motivation for the development of a new metadata model for nanoscience, details metadata implementation effort, specifically the ongoing work of metadata crosswalks between NFFA-EUROPE and EUDAT [8] (in Sect. 3.4), and presents a new refined version of the Common Vocabulary (in Appendix A) that underpins all metadata design and relates to other metadata artefacts that constitute the high-level metadata model. Also, this work outlines the identified challenges of metadata design and suggests directions for its further development (in Sect. 4).

2   Approach and Methodology

2.1   General Approach

The major purpose of any metadata is satisfying information needs of a certain community. "Community" should be understood in broad terms and includes machine agents, to ensure human-to-human, human-to-machine and machine-to-machine interoperability.

The information needs may be generic (common with other communities) or specific for a particular community. From the implementation point of view, the information needs should be expressed as clearly formulated Use Cases for the existing or proposed information and data management systems (IT platforms), so that the role of metadata in the data workflow can be clearly identified. A good metadata design should take into account user requirements and IT architecture, and in turn should feed considerations into the IT architecture. Figure 1 illustrates the approach taken in NFFA-EUROPE for the metadata design.

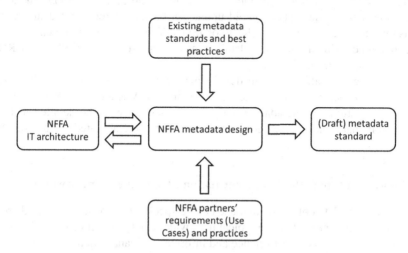

Fig. 1. Approach to NFFA-EUROPE metadata design.

Metadata can be considered a part of the *enterprise architecture* that includes both technological and organizational aspects of a loosely coupled virtual enterprise that the NFFA-EUROPE project is going to deliver for the European nanoscience community.

The metadata design then represents one of the pillars of the enterprise architecture design of the NFFA-EUROPE virtual enterprise, the other two pillars being business analysis and IT architecture design. Working on all three pillars should be mutually communicated and eventually aligned, which allows for the delivery of a quality enterprise architecture.

A good practice of information and data management adopted in the NFFA-EUROPE context is getting a good common understanding shared by the project partners about what actors (stakeholders), entities and relationships are most important in their domain and hence should be taken into account for the metadata design, and what are less important or too specific to be taken into common consideration. Through the iterative discussions in the project, we picked up the most relevant Roles and Responsibilities in the nanoscience domain, and mixed them up with the major Entities definitions that often constitute a basis of a structured formal knowledge representation (ontology) of a certain subject domain, but in our less ambitious case will form a basis of a reasonable metadata schema.

These discussions resulted in the Common Vocabulary (see Appendix A) which is a concise Body of Knowledge that describes information entities and relations between them that are most common in the project partners' experimental and data management environment. As a particular although again generic representation of this Body of Knowledge, we have described this in an Entity-Relationship (ER) diagram (see Sect. 3.2).

The Common Vocabulary and the ER diagram taken together with metadata groups and elements (see Sect. 3.1) constitute a generic metadata model and a baseline for all discussions about NFFA-EUROPE metadata. They are the basis for the detailed metadata model with the definition of metadata elements and relations among them. The detailed metadata model, when agreed upon, can be further represented in a certain serialisation format such as XML, RDF, or JSON. There is an early indication driven by technology considerations that a detailed master representation of NFFA-EUROPE metadata will be in JSON format.

The practice of iterative metadata development which we follow in NFFA-EUROPE has already got then a sound foundation – a Common Vocabulary, ER diagram and practical suggestions on metadata groups and elements – with the detailed metadata design and its particular (serialised) representations to be elaborated in later stages of the project.

2.2 Top-Down Input: Relevant Information Management Frameworks

The case for metadata collection and use can be specific to nanoscience, yet there are general information needs that are typical for a wide variety of users and that have been developed in other branches of science and information management.

One of the mature information design frameworks is Functional Requirements for Bibliographic Records (FRBR) [2] that considers four basic information needs (user tasks) in regards to information: "Find", "Identify", "Select" and "Obtain". The ultimate goal is of course getting the information resource, yet between searching for it and obtaining it, the resource should be identified as the one being sought, and selected as being useful for the user [1]. Each task may involve certain subtasks, e.g. selection may require checks on the resource context and on its relevance to the actual user's needs.

Another elaborated information design framework of relevance is the Reference Model for an Open Archival Information System (OAIS) [3], a widely-known functional model for long-term digital preservation. If expressed in terms of information practitioner needs (user tasks) similarly to FRBR, the OAIS basically deals with three categories of them: "Ingest (into archive)", "Manage (within archive)" and "Disseminate (from archive)". Each of these tasks may be complex and involve a number of interrelated subtasks, e.g. managing information in the archive may imply provenance and integrity checks, managing access to information, and administration/reporting.

Overall, the OAIS framework should be able to provide a good coverage of what NFFA-EUROPE needs to consider for sensible data collection, archiving and provision towards the end users (researchers in nanoscience), and the FRBR framework should be able to cover the end user needs for information retrieval. The respective areas of coverage and user categories relevant to NFFA-EUROPE are illustrated by the Table 1.

Table 1. Information management frameworks and their coverage of NFFA-EUROPE scope.

Framework (a source of best practices)	OAIS	FRBR
General use case	Data collection, management and dissemination	Data retrieval
User categories	Data archives administrators IT specialists	End users (nanoscience researchers)
Information needs (user tasks)	Ingest data Manage data Disseminate data	Find data Identify data Select data Obtain data

Being general in nature, OAIS and FRBR are still able to provide good recommendations for NFFA-EUROPE practices of information and data management. In particular, OAIS emphasizes the need of having a clear agreement between the data producer and the archive, and a clearly defined format for data exchange between them – so called Submission Information Package, whilst FRBR emphasizes the importance of having a clear identity for data assets.

2.3 Bottom-Up Input: Questionnaire Responses and Common Vocabulary

A questionnaire was used to collect the NFFA-EUROPE partners' responses about their data management practices and most popular data management solutions. The questionnaire inquired on the following aspects of data management in nano-facilities:

- Intensity of experiments and of resulting data flow
- Popular data formats
- Data catalogue software
- Data catalogue openness
- Data management policy
- Metadata standards for data catalogue
- Persistent identifiers for data
- User management platform
- Popular third-party databases and information systems

In total, seventeen responses out of the twenty project partners were received and reviewed. They showed very different levels of data management maturity. From the responses, the following priorities for metadata design were identified:

- One experiment to many samples and one sample to many data files relationships should be supported.
- A common set of metadata fields for data discoverability should be agreed upon, possibly based on an existing popular standards or recommendation for data discovery.
- User roles with different permissions for access to metadata should be developed. This means the metadata model will need to represent users as well as data.
- It is reasonable to develop a common data management policy for NFFA-EUROPE, or a set of policies with different flavours of access to data.
- Having links to external reference databases is valuable to ensure the high quality of metadata yet this will mean additional effort so should be de-scoped from the initial design of metadata.

In addition to the questionnaire where responses were collected from research offices or relevant research programme representatives, a common vocabulary of terms and definitions relevant to nanoscience data management was compiled and then refined by the IT teams of participating NFFA-EUROPE organizations (see Appendix A). The vocabulary contains commonly agreed terms with definitions; it serves as a basis for the design of information entities (groups of metadata elements) and contributes to the earlier mentioned NFFA-EUROPE "virtual enterprise" architecture.

2.4 Side Input: IT Architecture Considerations

As an additional consideration for principal metadata design, we used the draft of NFFA-EUROPE Data System Architecture that defines the outline design of the NFFA-EUROPE portal, which considered the generic use case of the same user performing a measurement on multiple facilities. Generic use cases when one user wants to access data produced by another user, or wants to release data into the public domain are currently not being considered. These may be considered in future, so should be taken into account within an extensible metadata design.

The draft architecture suggests that data should be harvested from individual facilities in a suitable "packaged" format, with METS [6] as a potential candidate as it supports the provision of descriptive, administrative, structural and file metadata. For the descriptive part of metadata, the purpose of having the data assets discoverable is emphasized in the draft architecture. For the administrative metadata, the importance of intellectual property information and information about the data source (provenance) is emphasized. For the structural metadata, having the information about the organization, perhaps structured in a hierarchical way, is suggested. For the file metadata, having the list of files that constitute a digital object (data asset) and having pointers to external metadata files are deemed most important.

After considering the draft of Data System Architecture, the conclusion was that we could take METS as "the role model" metadata standard that informs us about good practices of metadata design but we should not accept it as a default universal solution, as it does not cover all information needs of NFFA-EUROPE users. As to particular elements of metadata suggested by the Data System Architecture draft, the fields for capturing intellectual property information and provenance are easily most important ones as they affect the data assets reusability that should be one of the important outcomes of the NFFA-EUROPE project.

3   Implementation

3.1   Metadata Groups and Elements

The top-down, bottom up and side requirements resulted in the basic structure of the proposed metadata model that is illustrated by Fig. 2. This metadata structure generally reflects the data lifecycle in nano-science: first, an experiment is planned and conducted; it then results in some data assets (which can be measurements performed during nano-sample characterization, or controlled parameters of the sample physical production or computer simulation), then the archive that holds data assets should have its own operational requirements – again reflected in the respective section of metadata.

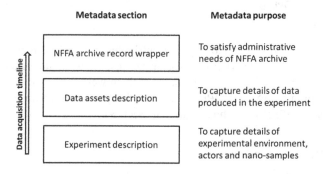

Fig. 2. Metadata groups of elements and their purpose.

The suggested metadata elements are presented as a matrix in Table 2 to make explicit the coverage of identified information entities (Common Vocabulary terms) and of earlier identified information needs (categories of them, see Sect. 2.2). Certain elements are in common with the Core Scientific Metadata Model [4] already in use in some of the facilities involved in the NFFA-EUROPE project.

Table 2. Metadata elements and information needs coverage.

Metadata section	Information entity	Ingest data	Manage data	Disseminate data	Find data	Identify data	Obtain data
Experiment description	Research user			Y	Y	Y	Y
	Instrument scientist	Y	Y				
	Project			Y	Y	Y	Y
	Proposal	Y	Y				
	Facility	Y	Y	Y	Y	Y	Y
	Instrument			Y	Y	Y	
	Experiment			Y	Y	Y	
	Sample			Y	Y	Y	
Data assets description	Data asset	Y	Y	Y	Y	Y	Y
	Raw data	Y	Y	Y	Y	Y	Y
	Analysed data	Y	Y	Y	Y	Y	Y
	Data analysis	Y	Y			Y	
	Data analysis Software	Y	Y			Y	
Archive record wrapper	Data archive	Y	Y				Y
	Data manager	Y	Y				Y
	Data policy	Y	Y				
	NFFA-EUROPE portal		Y		Y		

Mandatory and optional metadata fields (attributes) for each element were defined and shared amongst project participants for further discussion in the form of the project deliverable [5]. Some elements and attributes of them were further refined through the process of mapping NFFA-EUROPE metadata to the metadata scheme used in EUDAT B2SHARE service [9, 10] which is detailed in Sect. 3.4.

3.2   Entity-Relationship Diagram

As a basis for further, more detailed metadata design and as a contribution to the IT architecture design, the Entity-Relationship diagram presented by Fig. 3 has been agreed.

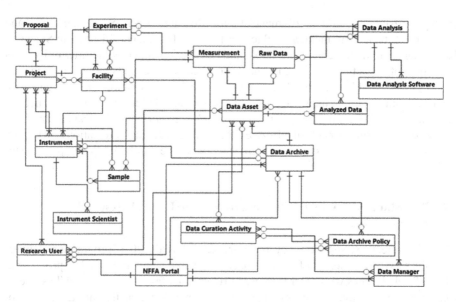

Fig. 3. NFFA-EUROPE metadata entity-relationship diagram.

This ER diagram has proven to be a useful tool for all discussions about NFFA-EUROPE metadata design; the entities in it relate to the terms in the Common Vocabulary (see Appendix A). The diagram allows at least three different perspectives: Research User-centric, Facility-centric, or Data Archive-centric, which reflects the natural data lifecycle in NFFA-EUROPE when the user first submits a research proposal, then conducts the actual (physical or/and computational) experiment, then NFFA-EUROPE takes the resulted data in custody.

3.3   Metadata Operational Recommendations

The metadata elements suggested are not all we need for having a successful metadata framework in NFFA-EUROPE. In addition, there should be established metadata management practices, ideally assisted by clear recommendations for NFFA-EUROPE partner organizations of how to assign and curate metadata.

For example, there are choices of how you aggregate data: let us say all data files for all samples measured in a particular Experiment can be assembled in one package, and then the package is given common descriptions such as Facility name, research User name, Data Policy etc. However, this may not suit actual data management practices or policies of certain Facilities, e.g. they may want to make a Sample rather than an Experiment a focal point of their metadata descriptions.

Another operational aspect important for the NFFA-EUROPE metadata scheme adoption by nanoscience community is actual levels of metadata that users and nano-facilities will be happy to provide when submitting research proposals and conducting (physical or computational) experiments. Initial evaluation performed using the research proposals submission system that is already in operation [11] has shown that it can provide satisfactory amount of metadata for Research User, Sample, Project and Proposal Entities. More metadata values for Facility, Instrument, Instrument Scientist, Experiment and Measurement entities should be supplied either by facilities or by users in the time of the actual experiment. The rest of metadata elements will be filled in with actual values by NFFA-EUROPE data portal. The population of metadata scheme with the actual values will be happening thus by various stakeholders and in stages that can be designated as "Research proposal submission" – "Experiment" – "Data archiving".

These operational aspects of NFFA-EUROPE metadata implementation will require further engagement and discussions with data practitioners in NFFA participant organizations.

3.4 Publishing NFFA-EUROPE Data Records in EUDAT Research Infrastructure

EUDAT project [8] supported by the European Horizon 2020 programme delivers common services in support of research data management and research data processing. EUDAT collaborates with other European projects that favour using the EUDAT services or software in place of development of their own functionally similar services or software.

NFFA-EUROPE have decided on the pilot use of EUDAT B2SHARE software platform [9, 10] to publish the data resulting from NFFA-EUROPE experiments in nano-facilities. The publication of NFFA-EUROPE data in B2SHARE will be subject to a data policy that is currently under the development in NFFA-EUROPE; in the meanwhile, there is a collaborative effort in NFFA-EUROPE and EUDAT to develop a metadata crosswalk from NFFA-EUROPE to EUDAT B2SHARE schema.

There is a common part of metadata schema in EUDAT B2SHARE that is universal across all communities who are using B2SHARE, and there is a community-specific part that B2SHARE platform can adopt as a template and then offer it for all individual researchers or institutions in the respective research domain – which will be nano-science in our case.

Both the universal part of NFFA-EUROPE metadata for B2SHARE and the community-specific part are first being discussed within NFFA-EUROPE, then with EUDAT representatives, to ensure semantic interoperability of metadata elements. The universal (community-unspecific) part of NFFA-EUROPE metadata crosswalk to B2SHARE schema is now fully agreed, and the crosswalk for the community-specific part of metadata is under development.

The actual data publishing from NFFA-EUROPE data portal to EUDAT B2SHARE instance will be performed using the B2SHARE API. We also foresee the situation when individual nano-science researchers or institutions will like to augment the B2SHARE instance (prepopulated with the automatically acquired data) with their own data uploaded via the B2SHARE user interface. Either a bundle of data and its metadata, or metadata only (with a reference to the corresponding data asset) can be uploaded in B2SHARE, that will give a proper flexibility for the nano-science researchers to share their data according to their local policies and personal preferences.

4   Identified Challenges and Further Developments

Apart from the clearly perceived need to develop, in addition to metadata schema, some operational recommendations for metadata curation (see Sect. 3.3), much needs to be done about better identity of metadata elements and values of their attributes.

For some information entities, having both an ID (which can be internal – specific to the facility or data management platform) and a PID (which should be universal) has been suggested: one of them intended for managing data in the NFFA-EUROPE software platform, and another for publishing the project outcomes beyond its boundary and lifespan.

It is the project's intention to get a registered URI for each metadata element – using PURL.ORG or similar services for managing namespaces and unique identifiers. The exact service and naming will be agreed through a dedicated discussion in the project. Unique URIs for metadata elements can constitute a basis for the further sharing of nanoscience data records as Linked Open Data, although the actual implementation of it is going to be beyond the NFFA-EUROPE scope.

Another addition likely to be required will be specifically designed fields for cross-linking metadata elements. As an example, Instrument may require a field, or a few, as a "foreign key" (which is only a metaphor, as the actual metadata representation may not be relational-based) to Facility; the same applies to a desirable link between Proposal and Research User, as well as to a number of other cases. The exact design of these fields dedicated to cross-linking of metadata elements will depend on the chosen format/syntax for metadata serialization: XML, RDF, JSON, or anything else.

We consider the necessity of introducing roles or types for certain metadata elements, up to the point of convergence of certain metadata elements into more universal ones supplied with a role or type attribute (a tag). Prime candidates for this would be Raw Data and Analysed Data elements, as both during and after the experiment, it may make sense to deal with "data continuum" where the data is assigned with appropriate tags depending on particular data collection, filtering or analysis steps.

Also, the detailed design of Data Asset has been postponed, as it will be heavily driven by the IT Architecture considerations and the pilot implementation of data portal, initially with only a few participating nano-facilities. A preliminary discussion suggested that METS could be a good metadata recommendation to model Data Asset, or to serve as a conceptual wrapper to the bespoke Data Assets modelling.

Certain considerations have been given to the notion of data processing workflows, although owing to the conceptual and technological complexity of workflows they are left beyond the metadata design in NFFA-EUROPE. Some suggestions of how one could model workflows, to a certain extent, by the means of the suggested NFFA-EUROPE data model can be found in Common Vocabulary (see Appendix A, specifically the definition of Data Analysis).

For Sample, there is a reserved metadata attribute for linking a brief record of it to a detailed one that is formed according to an existing standard. CODATA UDS [7] is considered a good candidate for a detailed and well-structured description of nano-science samples, so the current vision is just to rely upon a rich description of nano-samples offered by CODATA UDS if the NFFA-EUROPE ever identifies a need for a detailed samples description. The promotion of this or other suitable metadata standard for samples will be done then through the engagement effort across the project partners; this effort should be more of an operational nature rather than immediately related to the task of NFFA-EUROPE metadata design.

The Working Groups and Interest Groups of Research Data Alliance [12] are considering appropriate metadata frameworks for data sharing, both domain-focussed, e.g. dedicated to materials science, and cross-domain like those considering the best practices for persistent data identifiers. This is complementary to the approach of NFFA-EUROPE, and we foresee that this will be an appropriate forum for the continued metadata design for nanoscience.

5   Conclusion

The process of metadata development in NFFA-EUROPE so far has produced an agreed common approach with its mapping to the existing metadata frameworks and best practices. It has defined the common vocabulary, the structure of metadata groups and elements, the provisional list of mandatory and optional attributes, and the ER diagram that can be used both in metadata design and in IT architecture design. The high-level metadata model will be further refined through project work in NFFA-EUROPE and through discussions in the wider nanoscience community, with cooperating e-infrastructures like EUDAT and with relevant Research Data Alliance groups.

Acknowledgements.   This work is supported by Horizon 2020 NFFA-Europe project www.nffa.eu under grant agreement no. 654360.

Appendix A. Common Vocabulary for Nanoscience Data Management

This vocabulary is one of the components of the suggested high-level metadata model, along with the metadata groups and elements (see Sect. 3.1) and ER diagram (see Sect. 3.2) and hence as explained in Sect. 2.1 it is a contribution to the NFFA enterprise architecture, with a specific role of giving a common terminology for data practitioners in nanoscience. All the terms should be interpreted broadly with the inclusion of "in silico" experimental perspective, even if this is not explicitly mentioned. The vocabulary will be modified and expanded as necessary through further project works on metadata. **Research User.** A person, a group of them, or an institution (organization) who conduct Experiment on one or more nanoscience Facilities using one or more nanoscience Instruments to collect and analyse Raw Data, or is interested in data collected or analysed by other Research Users on the same or other Facilities. Research User may be assigned with a role, e.g. to designate the user as a principal investigator.
Instrument Scientist. A person or a group of them who manage a particular Instrument, or a set of them.

Project. An activity, or a series of activities performed by one or more Research Users on one or more Facilities using one or more Instruments for taking one or more Measurements of one or more Samples during one or more Experiments. Facility, Instrument, Measurement and Sample can refer to computer simulation environment. Project may involve one or more Proposals.

Proposal. An application of Research User for to perform a set of Experiments on one or more Facilities using one or more Instrument.

Facility. An institution (organization), or a division of it that operates one or more nanoscience Instruments for Research Users. For computer simulation, Facility may include hardware or/and software platform or/and services that allow to order and manage computational experiments (so that the software platform serves the purpose of managing software modules that can be considered virtual Instruments).

Instrument. Identifiable equipment (such as a device or a stand or a line) that allows conducting an independent nanoscience research, perhaps without involvement of other Instruments. Instrument is hosted by Facility and used by Research User. Instrument may be used for Sample production. Measurements conducted on Instrument result in Raw Data in the course of Experiment. Instrument can be in fact a software for computer simulation (a software module or/and a particular configuration of it).

Experiment. Identifiable activity with a clear start time and clear finish time conducted by Research User who uses Instrument to investigate or produce Sample and collects Raw Data about it. Experiment consists of (or includes – in case of Sample production) one or a series of Measurements and may also include one or a series of Data Analyses, potentially specific to Measurements. Experiment can be a computer simulation (computational experiment), or a combination of it with physical Measurements.

Measurement. The act of data collection for a Sample or a series of Samples during Experiment using a particular Instrument. Measurement can be a computer simulation, e.g. a particular run of a program using a particular model, configuration or input. Depending on a particular research context, Measurement may involve measuring the same sample under different conditions, or measuring different samples under the same conditions. Measurement is specific to Instrument: if one has to research the same Sample on a different Instrument it will imply a separate Measurement.

Sample. Identifiable piece of material with distinctive properties (structural, dimensional and others) exposed to Instrument during Experiment. Sample may stand for a model or configuration or data input (or any combination of these) in computer simulation.

Raw Data. Identifiable unit of data collected by Research User during Experiment. Raw Data is a result of Measurement. Unit of data is typically a data file but it can be potentially a data stream, or other form of data relevant in a particular data management context. Raw Data can be a result of computer experiment (simulation). Raw Data is always a part of Data Asset which may bear some semantics of what the data is and the origin/provenance of it.

Analysed Data. Identifiable unit of data which is a result of Raw Data processing obtained with the use of Data Analysis Software, typically after the end of Experiment. Unit of data is typically a data file but it can be potentially a data stream, or other form of data relevant in a particular data management context. Analysed Data may or may not be stored in the same Data Archive as Raw Data. Analysed Data can be a part of Data Asset which may bear some semantics of what the data is and the origin/provenance of it.

Data Asset. A combination of data units which can be Raw Data (including a result of computer simulation), Analysed Data, or Data Analyses (configurations or/and logs of Data Analyses execution). Depending on a particular data management context, Data Asset can be a dataset, a collection, or other form of data units organization. Data units remain identifiable within Data Asset. Data Asset allows capturing relationships between data units or/and their origin/provenance (e.g. corresponding Measurements or Data Analyses) or/and data curation operations performed on data units (e.g. checksum calculation). Data Asset may also serve as a "container" for different manifestations of the same data, e.g. for a collection of semantically equal data files in different formats. Data Asset can be used to express an accumulated result of Measurement (perhaps over multiple Samples).

Data Analysis. The identifiable action of processing Raw Data or/and Analysed Data, or a Data Asset with Data Analysis Software. Data Analysis can be thought of as something similar to Measurement – just input for it is not Sample but already collected data (raw or/and analysed or/and contextualized data collections/Data Assets). As Analysed Data can be a subject of Data Analysis, one can combine Data Analyses in chains or workflows. The definition of workflows and means of modelling them, however, is beyond the project scope, so no specific entities for workflows have been introduced in the metadata model; if someone wants to model workflows, the only means for that is currently Data Asset. Possible relation between Data Analysis and Data Asset is therefore twofold: on one hand, Data Analysis may use Data Assets as input; on the other hand, Data Asset may include Data Analyses configuration (or records of their execution).

Data Analysis Software. Software used for Raw Data analysis (that includes data rendering/visualization) and yields Analysed Data as an output. If software is used for simulation (computer experiment), is it considered Instrument and should be described as such.

Data Archive. An operational information system (repository) for Raw Data or/and Analysed Data on a certain Facility with certain rules and principles of data registration and management. Data Archive may or may not be used by Research User(s). Data Archive may include data storage solution (platform, component) and data catalogue solution (platform, component). Term "archive" should be interpreted broadly, i.e. it may be as simple as a file system, also the archive may not be supported by the Facility itself but by a certain third-party that Facility has an agreement with. Data Archive manages Data Assets according to Data Policy (which is perhaps specific to a particular type of Data Asset). Data Archive may be associated with a certain Facility or a group of them, or a certain Instrument or a group of them, or it may be run by a third-party where Facilities or Instruments are willing or obliged to supply their Data Assets (e.g. a discipline-wide or national archive). An example of third-party Data Archive not associated with a particular Facility is EUDAT B2SHARE. NFFA Portal may have one or more Data Archives as a back-end, or interoperate with them.

Data Policy. An identifiable expression of rules and regulations about data management in Data Archive (that includes data ingest) and about data sharing within and beyond Facility. Data Policy may be applicable to Raw Data or/and Analysed Data. Data Archive may have different Data Policies for different types of Data Assets. NFFA Portal (or its back-end Data Archive) may have one or more Data Policies, too.

Data Manager. Identifiable person, a group of them, an organizational unit, or a machine agent (software) who operate Data Archive on a certain Facility or in the third-party establishment that Facility or NFFA Portal have an agreement with. Having a clear identity and clear description of Data Manager is important for managing data harvesting (or federated data infrastructure) in NFFA Portal and resolving potential issues with Data Policies. It is also important for planning, performing and monitoring Data Curation Activities. Data Managers may have different roles; more than one role may be required by Data Archive or NFFA Portal, e.g. with different sets of permissions.

Data Curation Activity. An identifiable unit of work performed by Data Manager (in a certain role), or by a few of them. Examples of Data Curation Activity: data ingest, data integrity check, data transformation, restructuring or annotating data or collections of them. Data Curation Activity is performed on Data Assets according to Data Policies.

NFFA Portal. An IT service for nanoscience data discovery and sharing; the service may include one or more than one of: Graphical User Interface; Application Programming Interface; data ingestion and data publishing feeds; data sharing, data annotation and data analysis components. NFFA portal is used by Research Users and is underpinned by Data Archives in participating Facilities. Research Users may be registered with NFFA Portal. Data Archives of participant organizations may interact and interoperate with NFFA Portal – both technically and organizationally, e.g. by having Service Level Agreements for data supply in NFFA Portal.

References

1. Hider, P.: Information Resource Description: Creating and Managing Metadata. Facet Publishing, London (2012)
2. Functional Requirements for Bibliographic Records (FRBR). Final Report. http://archive.ifla.org/archive/VII/s13/frbr/
3. Reference Model for an Open Archival Information System (OAIS), Recommended Practice, CCSDS 650.0-M-2 (Magenta Book). Issue 2, June 2012. CCSDS (The Consultative Committee for Space Data Systems), Washington DC (2012)
4. The Core Scientific Metadata Model (CSMD). https://icatproject.org/user-documentation/csmd/
5. Draft metadata standard for nanoscience data. NFFA project deliverable D11.2. February 2016
6. METS: Metadata Encoding and Transmission Standard. http://www.loc.gov/standards/mets/
7. Uniform Description System for Materials on the Nanoscale, Version 2.0. CODATA-VAMAS Working Group on the Description of Nanomaterials (2016). doi:10.5281/zenodo.56720
8. EUDAT E-infrastructure project. http://www.eudat.eu/
9. EUDAT B2SHARE service. https://b2share.eudat.eu/
10. EUDAT B2SHARE user documentation. https://eudat.eu/services/userdoc/b2share
11. NFFA user guide. http://www.nffa.eu/apply

12. Research data alliance. https://www.rd-alliance.org/
13. Bunakov, V., Matthews, B., Griffin, T., Cozzini, S.: Metadata for nanoscience experiments. In: Selected Papers of the XVIII International Conference on Data Analytics & Management in Data Intensive Domains (DAMDID/RCDL 2016), CEUR Workshop Proceedings, vol. 1752, pp. 3–6 (2016)

Position Paper

Metrics and Rankings: Myths and Fallacies

Yannis Manolopoulos[1(✉)] and Dimitrios Katsaros[2]

[1] Department of Informatics, Aristotle University, 54124 Thessaloniki, Greece
manolopo@csd.auth.gr

[2] Department of Electrical and Computer Engineering, University of Thessaly, 38221 Volos, Greece
dkatsar@inf.uth.gr

Abstract. In this paper we provide an introduction to the field of Bibliometrics. In particular, first we briefly describe its beginning and its evolution; we mention the main research fora as well. Further we categorize metrics according to their entity scope: metrics for journals, conferences and authors. Several rankings have appeared based on such metrics. It is argued that these metrics and rankings should be treated with caution, in a light relative way and not in an absolute manner. Primarily, it is the human expertise that can rigorously evaluate the above entities.

Keywords: Scientometric indicators · Academic performance · Impact factor · CiteScore index · *h*-index · DORA · Rankings · Informetrics · Scientometrics

1 Introduction

The term *"Bibliometrics"* has been proposed by Alan Pritchard in 1969 [50]. According to Wikipedia, "Bibliometrics is statistical analysis of written publications, such as books or articles" [7]. A relevant field is *"Scientometrics"*, a term coined by Vasily Nalimov in 1969 [45]. According to Wikipedia, "Scientometrics is the study of measuring and analysing science, technology and innovation" [53]. Finally, *"Citation analysis"* is a fundamental tool for Bibliometrics and deals with the "examination of the frequency, patterns, and graphs of citations in documents" [14].

A milestone in the development of the field of Bibliometrics was the introduction of the *"Journal Impact Factor"* (IF) by Eugene Garfield in 1955 [25]. Garfield founded the Institute for Scientific Information (ISI) in 1964, which published the Science Citation Index and the Social Science Citation Index. ISI was acquired by Thomson Reuters in 1992.

For about four decades IF was the standard tool for academic evaluations. Despite the fact that it was proposed as a metric to evaluate journals' impact, it was used as a criterion to evaluate the quality of scholarly work by academicians and researchers as well. It was only in 2005 that Jorge Hirsch, a physicist, proposed the *h-index* as a simple and single number to evaluate the production and the impact of a researcher's work [32]. During the last 10–15 years the field has flourished and significant research has appeared in competitive journals and conferences.

L. Kalinichenko et al. (Eds.): DAMDID/RCDL 2016, CCIS 706, pp. 265–280, 2017.
DOI: 10.1007/978-3-319-57135-5_19

Based on these metrics, several rankings have appeared in the web, e.g., for journals, conferences and authors. On the other hand, university rankings appear in popular newspapers; actually, they are the beloved topic of journalists and politicians. University rankings are commercial artifacts of little scientific merit as they are based on arbitrary metrics (like the ones previously mentioned) and on other unjustified subjective criteria.

The purpose of this position paper is to explain that these metrics and rankings should be used with great skepticism. To a great extent, they shed light only to some particular facets of the entity in question (be it a journal, an author etc.); moreover, they are often contradictory to each other. The suggestion is to use this information with caution and pass it through an expert's filtration to come up with a scientific and objective evaluation.

The rest of the paper has the following structure. In the next section we provide more information about the Bibliometric/Scientometric community. Then, we introduce and annotate several metric notions for the evaluation of journals, conferences and authors. In Sect. 4 we mention assorted rankings and pinpoint their contradictions, limitations and fallacies. We devote Sect. 5 to discussing university rankings. Section 6 focuses on a recent negative phenomenon, that of selling publications or citations. In the last section we introduce the Leiden manifesto, an article that tries to put academic evaluations in an academic (not commercial, not mechanistic) framework. In addition, we introduce the main parts of the DORA report, and finally we state the morals of our analysis.

This paper is based on an earlier look on these topics [43]. In particular, the main augmentation parts in the current paper are the following ones: metrics by the Nature publishing group (Subsect. 3.3), moving ahead with citation-based metrics (Subsect. 3.6), altmetrics (Subsect. 3.7), science for sale (Sect. 6), and a large part of the final discussion (Sect. 7).

2   The Bibliometrics Community

During the last two decades a new research community was formed focusing on bibliometric and scientometric issues.

The research output of this community appears in specialized journals and conferences. In particular we note the following journals: (i) Journal of the Association for Information Science and Technology published by Wiley, (ii) Scientometrics by Springer, (iii) Journal of Informetrics by Elsevier, (iv) COLLNET Journal of Scientometrics and Information Management by Taylor and Francis, (v) Research Evaluation by Oxford Journals. Other less established outlets include: (i) Journal of Data and Information Science – formerly, Chinese Journal of Library and Information Science, (ii) Frontiers in Research Metrics and Analytics by Frontiers, (iii) Journal of Scientometric Research by Phcog.Net, (iv) Cybermetrics published by CINDOC and CSIC organizations in Spain, and (v) ISSI Newsletter by the ISSI society.

Two major annual conferences are running for more than a decade. For example, the International Conference of the International Society for Scientometrics and Informetrics (ISSI) is organizing its 16th event at Yuhan/China in October 2017, whereas the 12th International Conference on Webometrics, Informetrics, and Scientometrics (WIS) of the COLLNET community has been organized in December 2016 at Nancy/France.

In addition, assorted papers appear in other major outlets related to Artificial Intelligence, Data Mining, Information Retrieval, Software and Systems, Web Information Systems, etc.

Notably, there are several major databases with bibliographic data. Among others we mention: Google Scholar [28] and the tool Publish or Perish [51], which runs on top of Google Scholar, MAS (Microsoft Academic Search) [44], Scopus [55] by Elsevier and Web of Science [68] (previous known as ISI Web of Knowledge) by Thomson Reuters and DBLP (Data Bases and Logic Programming) [19] by the University of Trier.

3   The Spectrum of Metrics

3.1   Impact Factor

As mentioned earlier, IF is the first proposed metric aiming at evaluating the impact of journals. For a particular year and a particular journal, its IF is the average number of citations calculated for all the papers that appeared in this journal during the previous 2 years. More specifically, this is the 2-years IF as opposed to the 5-years IF, which has been proposed relatively recently as a more stable variant.

IF is a very simple and easily understood notion; it created a business, motivated researchers and publishers and was useful for academicians and librarians. However, IF has been criticized for several deficiencies. For instance,

- it is based mostly on journals in English,
- it considers only a fraction of the huge set of peer-reviewed journals,
- it did not take into account (until recently) conference proceedings, which play an important role in scholar communication in computer science for example,
- it fails to compare journals across disciplines,
- it is controlled by a private institution and not by a democratically formed scientific committee, etc.

On top of these remarks, studies suggest that citations are not *clean* and therefore the whole structure is weak [42]. Also, a recent book illustrates a huge number of flaws encountered in IF measurements [62].

IF can be easily manipulated by the journal's editors-in-chief, who are under pressure in the competitive journal market. For example, the editors-in-chief:

- may ask from authors to add extra references of the same journal,
- may invite/accept surveys as these articles attract more citations that regular papers, or
- may prefer to publish articles that seem to be well cited in the future.

Proposals to fight against IF's manipulation, include the *Reliability IF (RIF)* [39], which considers citation and the length of impact.

There is yet another very strong voice against the over-estimation of IF. Philip Campbell, Editor-in-Chief of the prestigious *Nature* journal, discovered that few papers make the difference and increase the IF of a specific journal. For example, the IF value of Nature for the year 2004 was 32.2. When Campbell analyzed Nature papers over the relevant period (i.e., 2002–2003), he found that 89% of the impact factor was generated by just 25% of the papers [12]. This is yet another argument against the use of IF to evaluate authors. That is, an author should not be proud just because he published an article in a journal with high IF; on the contrary he should be proud if his paper indeed contributed in this high IF value. However, it is well known that the distribution of the number of citations per paper is exponential [29]; therefore, most probably the number of citations of a paper per year will be less than half of the IF value.

In this category another score can be assigned: the *Eigenfactor* developed by Jevin West and Carl Bergstrom of the University of Washington [6]. As mentioned in Wikipedia: "The Eigenfactor score is influenced by the size of the journal, so that the score doubles when the journal doubles in size" [24]. Also, Eigenfactor score has been extended to evaluate the impact at the author's level.

More sophisticated metrics have been proposed, not only for reasons of elegance but also in the course of commercial competition as well.

3.2 Metrics by Elsevier Scopus

It is well-known that IF values range significantly from one field to another. There are differences in citation practices, in the lag time between publication and its future citation and in the particular focus of digital libraries. It has been reported that "the field of Mathematics has a weighted impact factor of IF = 0.56, whereas Molecular and Cell Biology has a weighted impact factor of 4.76 - an eight-fold difference" [4].

Scopus, the bibliometric research branch of Elsevier uses two important new metrics: *SNIP* (Source Normalized Impact per Paper) [59] and *SJR* (Scimago Journal Rank) [54]. Both take into account two important parameters.

SNIP has been proposed by the Leiden University Centre for Science and Technology Studies (CWTS) based on the Scopus database. According to SNIP citations are normalized by field to eliminate variations; IFs are high in certain fields and low in others. SNIP is a much more reliable indicator than the IF for comparing journals among disciplines. It is also less open to manipulation. Therefore, normalization is applied to put things in a relative framework and facilitate the comparison of journals of different fields.

On the other hand, SJR has been proposed by the SCImago research group from Consejo SCImago research group from the Consejo Superior de Investigaciones Científicas (CSIC), University of Granada, Extremadura, Carlos III (Madrid) and Alcalá de Henares. SJR indicates which journals are more likely to have articles cited by prestigious journals, not simply which journals are cited the most, adopting a reasoning similar to that of Pagerank's algorithm [11].

On December 8th 2016, Elsevier launched the *CiteScore index* to assess the quality of academic journals. It is very similar to IF, i.e., to score any journal in any given year, both tot up the citations received to documents that were published in previous years, and divide that by the total number of documents. However, CiteScore incorporates two major differences that results in quite diverse rankings with respect to those produced by the traditional IF; firstly it considers the "items" published during the last three years (instead of the last two years considered by IF), and secondly, CiteScore counts all documents as potentially citable, including editorials, letters to the editor, corrections and news items [65]. Despite the simplicity of the differences, they may make a huge difference in rankings. For instance, it has been observed that mathematics articles are slow in acquiring citations (it takes around five years to build up their counters). Also, including more items, especially those that are less cited, may have an adverse effect on the average. For instance, The Lancet, gets a 44 in IF, but a 7.7 in CiteScore, thus it is outside the top-200 in CiteScore.

These metrics have been put into practice as the reader can verify by visiting websites of journals published by Elsevier.

3.3 Metrics by the Nature Publishing Group

The Nature Publishing Group, publisher of some of the top-quality journals in various fields, made its entrance in the world of metrics in 2014 by introducing the *Nature index* [47]. Nature Index maintains a collection of author affiliations collected from articles published in a selected group of 68 high-quality scientific journals. This collection is aggregated and maintained by Nature Research. The Nature Index provides a proxy for high-quality research at the institutional, national and regional level. The Nature Index is updated monthly, and a 12-month rolling window of data is openly available.

There are three measures provided by the Nature Index. The *Article Count* (AC) is the simplest of them; we say that a country or institution has AC equal to 1, if the country or institution has one scientist that (co-)authored one article. Thus, with AC the same article can contribute to multiple countries or institutions. To remove this effect, the Nature Index provides the second measure, namely the Fractional *Count* (FC). The total FC per article is 1, and it is split equally among the co-authors; if some authors have multiple affiliations, then the share is split equally among them. Finally, the *Weighted Fractional Count* (WFC) applies a weighting scheme to properly adjust the FC to the overrepresentation of articles from astronomy and astrophysics, because the four journals (out of the 68) in these disciplines account for about 50% of all the articles in the journals of these fields.

Various objections concern the validity of the Nature Index, such the identity and the way the 68 journals were selected, the citation counting per author, and the weighting scheme, and so on. Nevertheless, this index carries the authoritativeness of its promoter.

3.4 Metrics for Conferences

Conferences are not treated in a uniform way from one discipline to another and even from subfield to subfield. For example, there are conferences where only abstracts are submitted, accepted and published in a booklet, whereas there are other conferences where full papers are submitted and reviewed by ~3 referees in the same manner as journals. Apparently, the latter articles can be treated as first class publications, in particular if the acceptance ratio is as high as 1 out 5, or 1 out of 10 as it happens in several prestigious conferences (such as WWW, SIGMOD, SIGKDD, VLDB, IJCAI, etc.).

Thus, an easy metric to evaluate the quality of a conference is the acceptance ratio (i.e. number of accepted vs. number of submitted papers). Several publishing houses (e.g. Springer) specify that the acceptance ratio should be <33%. It is a common practice to report such numbers in the foreword of conference proceedings.

Several websites collect data about the acceptance ratios of conferences of several fields such as Theoretical CS, Computer Networks, Software Engineering etc. [1–3]. In general, there is a trend towards events with stricter acceptance policies. The humoristic study of [18] seriously deconstructs such approaches.

Apart from the acceptance ratio, an effort to quantify the impact of conferences has been first initiated by Citeseer, a website and service co-created by Lee Giles, Steve Lawrence and Kurt Bollacker at NEC Research Institute [26]. In particular, using its own datasets Citeseer calculates the IF of a rich set journals and conferences in a unique list. Nowadays, Citeseer is partially maintained by Lee Giles at the Pennsylvania State University [15]; practically, it has been surpassed by Google Scholar.

3.5 Metrics for Authors

In 1985, Jorge Hirsch, a physicist at UCSD, invented the notion of the *h-index* [32]. According to Wikipedia: "a scholar with an index of h has published h papers each of which has been cited in other papers at least h times" [34]. Thus, the *h*-index illustrates both the production and the impact of a researcher's work. It is not just a single number but a 2-dimensional number. *h*-index was a breakthrough; it was a brand new notion that broke the monopoly of IF in academic evaluations.

A propos, it came up that the *h*-index was just a re-invention of a similar metric. Arthur Eddington [23], an English astronomer, physicist, and mathematician of the early 20th century, was an enthusiastic bicycler. In the context of cycling, Eddington's number is the maximum number E such that the cyclist has cycled E miles on E days. Eddington's own E-number was 84 [22].

Although a breakthrough notion, the *h*-index received some criticism when it was put in practice. In particular, the following issues were brought up:

- it does not consider peculiarities of each specific field,
- it does not consider the order of an author in the list of authors,
- it has a reduced discriminative power as it is an integer number,
- it can be manipulated with self-citations, which cannot be revealed in Google Scholar,
- it has a correlation with the number of the author's publications,
- it constantly increases with time and cannot show the progress or stagnation of an author.

Soon after the invention of the h-index, the field of Bibliometrics flourished and a lot of variants were proposed. The following is only a partial list: g-index, a-index, h(2)-index, hg-index, q^2-index, r-index, ar-index, m-quotient, k-index, f-index, m-index, h_w-index, hm-index, h_{rat}-index, v-index, e-index, π-index, RC-index, CC-index, ch-index, n-index, p-index, w-index, and so on and so forth [35]. The present author's team proposed the following 3 variants: contemporary *h*-index, trend *h*-index, normalized *h*-index [57]. After this flood of variants, several such studies were reported aiming at analyzing, comparing and categorizing the multiplicity of indicators [8, 10, 69, 70]. Two interesting studies are

[48, 64], where *h*-index is criticized as easily manipulated by authors in an effort to improve their metrics.

In passing, there have been efforts in studying and devising metrics for the whole citation curve of the works by an author, as a metric supplementary to the *h*-index. In this direction, we proposed two new metrics: the perfectionism index [58] and the fractal dimension [27] to penalize long tails and to dissuade authors from writing papers of low value.

The books by Nikolay Vitanov [67] and Roberto Todeschini, Alberto Baccini [61] give extensive insight into these metrics for authors. In addition, in Publish or Perish [51], a website (maintained by Harzing) and book [30], the most common of these variants have been implemented. Finally, Matlab includes several such implementations as well.

3.6 Moving Ahead with Citation-Based Metrics

One of the major obstacles, for any citation-based metric is the radically different citation patterns from discipline to disciple, which causes problems in comparing journals, persons from different disciplines, and calls for an effective normalization method. To alleviate this problem, the *Relative Citation Ratio* (RCR) [37] was proposed recently; it is a field-normalized metric that shows the citation impact of an article relative to the average NIH-funded paper. It is accompanied by a free software for its calculation, namely iCite, The normalization procedure is done by using each article's co-citation network to field- and time-normalize the number of citations it has received; this topically linked group of articles is used to derive an expected citation rate, which serves as the ratio's denominator. The article citation rate (ACR) is used as the numerator. The basic idea behind RCR is clever, however the study reported in [9], establishes that the RCR correlates highly with established field-normalized indicators, but the correlation between RCR and peer assessments is only low to medium.

The skewness in citation distributions is a universal law, and thus makes metrics stemming from averages to misrepresent the citation curve. So, journals, such as those published by the Royal Society and EMBO Press, already publicize citation distribution [13]. Citation distributions are more relevant than impact factors for high-stakes decisions, such as hiring and promotion, but they can be useful for researchers who are trying to decide which among a pile of papers to read. Protocols for the publication of whole citation distributions appeared recently in literature [41], which can reveal the full extent of the skew of distributions and variation in citations received by published papers.

3.7 Altmetrics: The Alternative to Citation-Based Metrics

The term *altmetrics* was proposed in 2010 [49] as an alternative to article level metrics. Usually, altmetrics are thought of as metrics about articles, but they can be gracefully applied to persons, publication fora, presentations, videos, source code repositories, Web pages, etc. A classification of altmetrics was proposed by ImpactStory in September 2012 [38]:

- Viewed: HTML views and downloads.
- Discussed: journal comments, science blogs, Wikipedia, Twitter, Facebook and other social media.
- Saved: Mendeley, CiteULike and other bookmarking services.
- Cited: citations in the scholarly literature, tracked by Web of Science, Scopus, CrossRef and others.
- Recommended: for example used by F1000Prime.

Altmetrics are in principle more difficult to standardize compared to standard impact measures such as citations. One example is the number of tweets linking to a paper where the number can vary widely depending on how the tweets are collected. Like other metrics, altmetrics are prone to manipulation, by self-citation, gaming, and other mechanisms to boost one's apparent impact. For instance, altmetrics can be gamed in the following ways: likes and mentions can be bought.

4   The Spectrum of Rankings

Ranking is a popular game in academic environments. One can easily find rankings about authors, journals, conferences, and universities as well. Here, we comment on some interesting rankings drawn from several websites. In particular, we will comment on university rankings in the next section.

DBLP website [19] is maintained by Michael Ley at the University of Trier. As of May 2016, its dataset contains more than 1.7 million authors and 3.5 million articles. Based on this dataset, DBLP posts a list of prolific authors in terms of publications of all sorts, e.g. journal and conference papers, books, chapters in books etc. It is interesting to note that Vincent Poor, a researcher at Princeton University, is the most productive person in this ranking with an outcome of 1348 publication (as of 21/10/2016) [20]. Another ranking with the same dataset ranks authors according to the average production per year. Vincent Poor can be found in the 19th position in this ranking (as of 21/10/2016) [21].

Jens Palsberg of UCLA maintains a website where, by using the DBLP datasets, a list of authors ordered according to decreasing h-index is produced. First name in this list is Herbert Simon, a Professor at CMU, Nobel Laureate, Turing Award recipient, ACM Fellow; his h-index is 164. In this list, Vincent Poor appears with h-index equal to 70 [36].

MAS provides a variety of rankings using a dataset of 80 million of articles [44]. For example, it provides two ranked lists of authors according to productivity and according to impact. When the whole dataset is taken into account, then in terms of the number of publications Scott Shenker is 1st, Ian Foster is 2nd and Hector Garcia-Molina is 3rd. According to the number of citations Ian Foster is 1st, Ronald Rivest is 2nd and Scott Shenker is 3rd. Other ranking can be produced by limiting the time window during the last 5 or 10 years. For instance, Vincent Poor is ranked 2nd in terms of productivity for the period of the last 10 years.

Similarly, MAS provides rankings of conferences according to the number of publications or citations for certain periods (i.e., 5 years, 10 years, or the whole dataset). Steadily, INFOCOM, SIGGRAPH, CVPR, ICRA, ICASSP appear at the top.

In an analogous manner, MAS provides rankings for journals. When considering the whole data set, the top journals are CACM, PAMI and TIT. During the last 5 years, new fields came up and, thus, new journals gained acceptance: see for example Expert Systems with Applications and Applied Soft Computing. It is important to notice that these rankings use raw numbers, i.e. without any normalization. However, they show trends in science with time.

The above paragraphs show that there are several kinds of ranking, each with a different emphasis and as such they should be treated with caution. Another example of misuse of rankings concerns the classification of journals and conferences. CORE is an Australian website/service, where journals and conferences are divided in 5 categories as illustrated in the following table [17]. Numbers show the percentages of journals or conferences at their *corresponding* category. Similar categorizations exist in other websites. Even though it is not transparent how the percentages were calculated and the rankings are based on somewhat arbitrary listings and categorizations, such rankings have great acceptance and in several instances state funding may be based on them.

	A*	A	B	C	Other
Journals	7%	17%	27%	46%	3%
Conferences	4%	14%	26%	51%	5%

We give another example where caution is needed. We present two tables. The first table contains data from Aminer [5], which runs on top of DBLP. This table shows the top-10 outlets for "database and data mining" sorted by the $h5$-index, a variation of h-index for journals. $h5$ is the largest number h such that h articles published in 2011–2015 have at least h citations each. In an analogous manner, the following table gives the top-10 outlets for "Database and Data Mining" sorted by $h5$ according to Google Scholar for "Database and Information System".

1	WWW conference	66
2	Information Sciences	62
3	ACM KDD	56
4	IEEE TKDE	53
5	ACM WSDM conference	50
6	JASIST	47
7	ACM SIGIR	42
8	IEEE ICDE conference	40
9	ACM CIKM conference	38
10	IEEE ICDM conference	33

1	WWW conference	74
2	VLDB conference	67
3	IEEE TKDE	66
4	arXiv Social & Infor. Networks (cs. SI)	66
5	ACM SIGMOD conference	65
6	arXiv Databases (cs DB)	61
7	ICWSM (weblog) conference	60
8	ICWSM (web) conference	59
9	ACM WSDM conference	58
10	IEEE ICDE conference	52

We note that the two lists have only 5 items in common, and in different order. At first, one might think that the two lists were not comparable since they were produced by querying different key-words. However, since the first table contains outlets related to Information Systems, whereas the second one contains outlets related to Data Mining the two lists are indeed comparable. This example illustrated that the adoption of a ranking versus another is a subjective matter.

5   University Rankings

Nowadays education is considered as a product/service and, thus, there is a growing financial interest in this global market. Universities try to improve their position in the world arena. Thus, university rankings try to satisfy the need of universities for visibility. These ranking are a popular topic for journalists and, therefore, for politicians as well. However, beforehand we claim that there is little scientific merit in these rankings.

Some rankings are widely-known from mass media. We mention alphabetically the most commercial ones:

- Academic Ranking of World Universities (Shanghai or ARWU),
- QS World University Rankings (QS),
- Times Higher Education World University Rankings (THE).

Other rankings originate from academic research teams, such as:

- Leiden Ranking,
- Wikipedia Ranking of World Universities,
- Professional Ranking of World Universities (École Nationale Supérieure des Mines de Paris),
- SCImago Institutions Ranking,
- University Ranking by Academic Performance (Middle East Technical University),
- Webometrics (Spanish National Research Council),
- Wuhan University.

A full list of such rankings exists at Wikipedia [63].

University rankings are intensively criticized for a number of reasons. For example:

- All rankings are based on a number of subjective criteria.
- In all cases, the choice of each particular criterion and its weight are arbitrary.
- To a great extent, these criteria are correlated.
- Evaluation for some criteria is based on surveys, e.g. "academic reputation" or "employer reputation" by QS, which count for 50% of the total weight. The same holds for the "reputation survey" by THE, which counts for 17.9% or 19.5% or 25.3%, if the examined institution is a medical, an engineering or an arts/humanities school, respectively. Such surveys are totally not-transparent.
- THE devotes a 7.5% of the total weight for the international outlook, subcategorized into "ratio of international to domestic staff", "international co-authorship" and "ratio of international to domestic students". In the same way, QS considers "international student ratio" and "international staff ratio" with a special weight of 5% + 5%. Clearly, such criteria favor Anglo-Saxon universities.
- The number of publications and the number of citations (without normalization) favor big universities; this is probably a reason for a general trend in merging universities in Europe.
- No ranking considers whether a university is an old or a new institution, big or small, a technical university or a liberal arts one, etc. Thus, different entities are compared.
- In general, ranking results are not reproducible, an absolutely necessary condition to accept an evaluation as methodologically reliable.
- QS adopts the *h*-index at a higher level, i.e. not at the author's level but for a group of academicians. This is beyond the fair use of the original idea by Hirsch since it does not consider the size of the examined institution, neither has it performed any normalization.
- The rankings exert influence on researchers to submit papers to "prestigious" journals (e.g. Nature, Science). Since such journals follow particular policies as to what is in fashion researchers may not work on what they truly think is worthwhile but according to external/political criteria acting as sirens [56].
- Finally, and probably the most important point of this criticism is that university rankings are misleading proportionally to the degree that they are based on (a) collections of citations from English-language digital libraries, (b) erroneous collections of citations, (c) IF calculations, which ignore whole statistical distributions of a single number, (d) higher level *h*-index calculations, which are conceptually wrong. In other words, "garbage in, garbage out".

All rankings are not equally unacceptable. Several independent studies agree that ARWU is probably the most reliable in comparison to other commercial rankings [40], whereas QS is the most criticized ranking. On the other hand, between the rankings originated from academic institutions, Leiden is considered as the most reliable as it stems from a strong research team with significant academic reputation and tradition in the field of Bibliometrics/Scientometrics. On the other hand, the ranking of Webometrics is criticized for the adoption of non-academic criteria, such as the number of web pages and files and their visibility and impact according to the number of received inlinks.

Based on the above discussion, one can understand why the question "science or quackery" arises [52]. In a recent note by Moshe Vardi, Editor-in-Chief of CACM and professor with Rice University, same skepticism was reported [66]. Moreover, some state authorities are critical against these methodologies [46]. However, it is sad that rankings are "*here to stay*" because strong financial interests worldwide support such approaches.

At this point, we mention a very useful website which, based on the DBLP dataset, ranks American CS departments in terms of the number of faculty and the number of publications in selected fora, by picking certain CS subfields [16].

6 Science for Sale

There is yet another problem with the sole existence of any research impact/productivity indicator. Scholars in their struggle to increase their personal ranking according to such indicators, e.g., to publish articles in journals with high IF, to increase citation numbers, resort to the worst possible practice, for instance, to buy authorship! It has been reported in [33] that this highly unethical act is far beyond an isolated event. It is a developing market involving "shady agencies, corrupt scientists, and compromised editors", where the prices vary depending on whether the buyer wishes to be the first/primary author, or merely a coauthor. Similar unethical acts have been documented in the context of citation buying [60].

7 Discussion and Morals

The intention of this position paper is the following. Bibliometrics is a scientific field supported by a strong research community. Although the term is not new, during the last years there is an intense research in the area due to the web and open/linked data.

The outcome of Bibliometrics is most often misused by mass media and journalists, state authorities and politicians, and even in the academic world. Criticism has been expressed for several metrics and rankings, not without a reason.

In 2015, a paper was published in Nature entitled: "The Leiden Manifesto for research metrics". More specifically, the paper states 10 principles to guide research evaluation [31]. We repeat them here in a condensed style:

1. Quantitative evaluation should support qualitative, expert assessment.
2. Measure performance against the research missions of the institution, group or researcher.
3. Protect excellence in locally relevant research.
4. Keep data collection and analytical processes open, transparent and simple.
5. Allow those evaluated to verify data and analysis.
6. Account for variation by field in publication and citation practices.
7. Base assessment of individual researchers on a qualitative judgment of their portfolio.
8. Avoid misplaced concreteness and false precision.
9. Recognize the systemic effects of assessment and indicators.
10. Scrutinize indicators regularly and update them.

Probably, the last principle is the most important. Since it is easy for humans to cleverly adapt to external rules and to try to get the most benefit out of them, the Bibliometrics community has to devise and promote new metrics for adoption by academia and others.

At this point, it is worth mentioning the San Francisco Declaration of Research Assessment (DORA) [22], which was initiated by the American Society for Cell Biology (ASCB) together with editors and publishers of scholarly journals. They altogether recognize the need to improve the ways in which the outputs of research are evaluated. The group developed a set of recommendations, known as the San Francisco Declaration on Research Assessment. In summary, they published recommendations for four "bodies":

- For funding agencies:
  - To clarify the criteria used in evaluating the scientific output of grant applicants and state firmly that an article's content is much more important than metrics or the identity of the journal in which the article was published.
  - To account also for the significance and impact of collected datasets and developed software, apart from the published articles, and consider a wide range of impact metrics, including also qualitative indicators such as influence on decision making and practice.
- For institutions:
  - To be explicit about the criteria used to make decisions about hiring, tenure, and promotion, and state firmly that an article's content is much more important than metrics or the identity of the journal in which the article was published.
  - For the purposes of research assessment, to account for the significance and impact of collected datasets and developed software, apart from the published articles, and consider a wide range of impact metrics, including also qualitative indicators such as influence decision making and practice.
- For publishers:
  - To avoid giving too much emphasis on the IF, or at least to present IF as a member of a set of journal-related metrics (e.g., EigenFactor, SCImago, h-index, etc.) that provide a multidimensional perspective on journal performance.
  - To make available several measures concerning article impact so as to push toward assessment based on the content of an article rather than metrics of the journal in which it was published.
  - To enforce at the extent possible responsible authorship practices, and the clear statement about the specific contributions of each author.
  - To remove all reuse limitations on reference lists and make them available under the Creative Commons Public Domain Dedication.
  - To remove the constraints on the number of references in research articles.
- For organizations that supply metrics:
  - To be open by providing data and methods used to calculate all measures.
  - To provide the data under a license that allows for its unrestricted reuse.
  - To be clear that manipulation of metrics is unacceptable, and that any manipulation of them will have severe consequences.
  - To account for the variation in article types.
- For researchers:

- To make assessments involving funding, hiring, promoting based on true scientific contributions rather than publication metrics.
- To cite the articles where the research originated instead of reviews articles.
- To use several article measures to evaluate the impact of articles.
- To challenge research assessment practices that relies only on IFs.

Finally, we close this paper with a proposed personal list of do's and don'ts.

1. Do not evaluate researchers based on the number of publications or the IF of the journals they appeared.
2. Evaluate researchers with their h-index and variants (resolution according to competition).
3. To further evaluate researchers, focus on the whole citation curve and its tail in particular (relevant metrics: perfectionism index and fractal dimension).
4. Do not evaluate journals based on their IF.
5. Evaluate journals with the SCIMAGO and EIGENFACTOR scores as they are robust and normalized.
6. Further, ignore journal metrics and choose to work on the topics that inspire you.
7. Metrics are not panaceas; metrics should change periodically.
8. Do not get obsessed with contradictory rankings for authors, journals and conferences.
9. Ignore university rankings; they are non-scientific, non-repeatable, commercial, and unreliable.
10. Follow your heart and research what attracts and stimulates you.

Acknowledgments. Thanks are due to our ex and present students and colleagues. Many of the ideas expressed in this article are the outcome of research performed during the last 15 years. In particular, we would like to thank Eleftherios Angelis, Nick Bassiliades, Antonia Gogoglou, Vassilios Matsoukas, Antonios Sidiropoulos and Theodora Tsikrika.

References

1. Acceptance Ratio of Networking Conferences: https://www.cs.ucsb.edu/~almeroth/conf/stats
2. Acceptance Ration of SW Engineering Conferences: http://taoxie.cs.illinois.edu/seconferences.htm
3. Acceptance Ratio of TCS Conferences: http://www.lamsade.dauphine.fr/~sikora/ratio/confs.php
4. Althouse, B., West, J., Bergstrom, T., Bergstrom, C.: Differences in impact factor across fields and over time. J. Am. Soc. Inf. Sci. Technol. **60**(1), 27–34 (2009)
5. Aminer: https://aminer.org/ranks/conf
6. Bergstrom, C.T., West, J.D., Wiseman, M.A.: The Eigenfactor metrics. J. Neurosci. **28**(45), 11433–11434 (2008)
7. Bibliometrics: https://en.wikipedia.org/wiki/Bibliometrics
8. Bollen, J., van de Sompel, H., Hagberg, A., Chute, R.: A principal component analysis of 39 scientific impact measures. PLoS ONE **4**, e6022 (2009)
9. Bornmann, L., Haunschild, R.: Relative Citation Ratio (RCR): an empirical attempt to study a new field-normalized bibliometric indicator, J. Assoc. Inf. Sci. Technol. (2017, to appear)

10. Bornmann, L., Mutz, R., Hug, S., Daniel, H.D.: A multilevel meta-analysis of studies reporting correlations between the *h*-index and 37 different *h*-index variants. J. Inform. **5**(3), 346–359 (2011)

11. Brin, S., Page, L.: The anatomy of a large-scale hypertextual web search engine. Comput. Netw. ISDN Syst. **30**(1–7), 107–117 (1998)

12. Campbell, P.: Escape from the impact factor. Ethics Sci. Environ. Politics **8**, 5–7 (2008)

13. Callaway, E.: Publishing elite turns against impact factor. Nature **535**, 210–211 (2016)

14. Citation Analysis: https://en.wikipedia.org/wiki/Citation_analysis

15. CiteSeer Digital Library: http://citeseerx.ist.psu.edu/index

16. Computer Science Ranking: http://csrankings.org

17. Computing Research and Evaluation (CORE): http://www.core.edu.au

18. Cormode, G., Czumaj, A., Muthukrishnan S.: How to increase the acceptance ratios of top conferences? http://www.cs.rutgers.edu/~muthu/ccmfun.pdf

19. DBLP: http://dblp.uni-trier.de

20. DBLP: Prolific authors. http://dblp.uni-trier.de/statistics/prolific1

21. DBLP: Prolific authors per year. http://dblp.l3s.de/browse.php?browse=mostProlificAuthorsPerYear

22. DORA: http://www.ascb.org/dora/

23. Eddington Arthur: https://en.wikipedia.org/wiki/Arthur_Eddington

24. Eigenfactor Metric: https://en.wikipedia.org/wiki/Eigenfactor

25. Garfield, E.: Citation indexes for science: a new dimension in documentation through association of ideas. Science **122**, 108–111 (1955)

26. Giles, C.L., Bollacker, K., Lawrence, S.: CiteSeer: an automatic citation indexing system. In: Proceedings 3rd ACM Conference on Digital Libraries, pp. 89–98 (1998)

27. Gogoglou, A., Sidiropoulos, A., Katsaros, D., Manolopoulos, Y.: Quantifying an individual's scientific output using the fractal dimension of the whole citation curve, In: Proceedings 12th International Conference on Webometrics, Informetrics & Scientometrics (WIS), Nancy (2016)

28. Google Scholar: http://www.scholar.google.com

29. Gupta, H., Campanha, J., Pesce, R.: Power-law distributions for the citation index of scientific publications and scientists. Braz. J. Phys. **35**(4a), 981–986 (2005)

30. Harzing, A.W.: Publish or Perish, Tarma Software Research (2010)

31. Hicks, D., Wouters, P., Waltman, L., de Rijke, S., Rafols, I.: The Leiden Manifesto for research metrics. Nature **520**(7548), 429–431 (2015)

32. Hirsch, J.E.: An index to quantify an individual's scientific research output. Proc. Natl. Acad. Sci. **102**(46), 16569–16572 (2005)

33. Hvistendahl, M.: China's publication bazaar. Science **342**(6162), 1035–1039 (2013)

34. *h*-index: https://en.wikipedia.org/wiki/H-index

35. *h*-index Variants: http://sci2s.ugr.es/hindex

36. *h*-index for CS Scientists: http://web.cs.ucla.edu/~palsberg/h-number.html

37. Hutchins, B.I., Yuan, X., Anderson, J.M., Santangelo, G.M.: Relative Citation Ratio (RCR): a new metric that uses citation rates to measure influence at the article level. PLoS Biol. **14**(9), e1002541 (2016)

38. ImpactStory Blog. A new framework for altmetrics (2012)

39. Kuo, W., Rupe, J.: R-impact factor: reliability-based citation impact factor. IEEE Trans. Reliab. **56**(3), 366–367 (2007)

40. Lages, J., Patt, A., Shepelyansky, D.: Wikipedia Ranking of World Universities (2016). Arxiv https://arxiv.org/abs/1511.09021

41. Lariviere, V., Kiermer, V., MacCallum, C.J., McNutt, M., Patterson, M., Pulverer, B., Swaminathan, S., Taylor, S., Curry, S.: A simple proposal for the publication of journal citation distributions, Technical report. http://dx.doi.org/10.1101/062109

42. Lee, D., Kang, J., Mitra, P., Giles, L., On, B.W.: Are your citations clean? Commun. ACM **50**(12), 33–38 (2007)

43. Manolopoulos, Y.: On the value and use of metrics and rankings: a position paper. In: Selected Papers of the 18th International Conference on Data Analytics & Management in Data Intensive Domains (DAMDID 2016), vol. 1752, pp. 133–139. CEUR Workshop Proceedings (2016)

44. Microsoft Academic Search: http://academic.research.microsoft.com

45. Nalimov, V., Mul'chenko, Z.M.: Naukometriya, the study of the development of science as an information process in Russian, p. 191. Nauka, Moscow (1969)

46. Norwegian   Universities: http://www.universityworldnews.com/article.php? story=20140918170926438

47. Nature Publishing Group: A guide to the nature index. Nature **515**(7526), S94 (2014)

48. Piazza, R.: On house renovation and co-authoring – tricks of the trade to boost your h-index. Europhys. News **46**(1), 19–22 (2015)

49. Priem, J., Taraborelli, D., Groth, P., Neylon, C.: Altmetrics: a manifesto. altmetrics.org

50. Pritchard, A.: Statistical bibliography or bibliometrics? J. Doc. **25**(4), 348–349 (1969)

51. Publish or Perish: http://www.harzing.com/pop.htm

52. Science or Quackery: https://www.aspeninstitute.it/aspenia-online/article/international-university-rankings-science-or-quackery

53. Scientometrics: https://en.wikipedia.org/wiki/Scientometrics

54. SCIMAGO: http://www.scimagojr.com/

55. Scopus: http://www.scopus.com

56. Schekman, R.: https://www.theguardian.com/science/2013/dec/09/nobel-winner-boycott-science-journals

57. Sidiropoulos, A., Katsaros, D., Manolopoulos, Y.: Generalized Hirsch h-index for disclosing latent facts in citation networks. Scientometrics **72**(2), 253–280 (2007)

58. Sidiropoulos, A., Katsaros, D., Manolopoulos, Y.: Ranking and identifying influential scientists vs. mass producers by the perfectionism index. Scientometrics **103**(1), 1–31 (2015)

59. SNIP: http://www.journalindicators.com

60. The Daily Californian: http://www.dailycal.org/2014/12/05/citations-sale/

61. Todeschini, R., Baccini, A.: Handbook of Bibliometric Indicators. Wiley (2016)

62. Tüür-Fröhlich, T.: The Non-trivial Effects of Trivial Errors in Scientific Communication and Evaluation. Verlag Werner Hülsbusch, Glückstadt (2016)

63. University Rankings: https://en.wikipedia.org/wiki/College_and_university_rankings

64. Van Bevern, R., Komusiewicz, C., Niedermeier, R., Sorge, M., Walsh, T.: H-index manipulation by merging articles: models, theory and experiments. Artif. Intell. **240**, 19–35 (2016)

65. van Noorden, R.: Impact factor gets heavyweight rival: citeScore uses larger database and gets different results. Nature **540**, 325–326 (2016)

66. Vardi, M.: Academic rankings considered harmful! Commun. ACM **59**(9), 5 (2016)

67. Vitanov, N.: Science Dynamics and Research Production. Springer, Cham (2016)

68. Web of Science: http://ipscience.thomsonreuters.com

69. Wildgaard, L., Schneider, J.W., Larsen, B.: A review of the characteristics of 108 author-level bibliometric indicators. Scientometrics **101**, 125–158 (2014)

70. Yan, Z., Wu, Q., Li, X.: Do Hirsch-type indices behave the same in assessing single publications? An empirical study of 29 bibliometric indicators. Scientometrics **109**(3), 1815–1833 (2016)

Author Index

Printed in the United States
By Bookmasters

Printed in the United States
By Bookmasters